DARWIN AND THE BARNACLE

DARWIN
AND THE BARNACLE

*

Rebecca Stott

W. W. NORTON & COMPANY
New York • London

First published in Great Britain in 2003 by Faber and Faber Limited
Copyright © 2003 by Rebecca Stott
First American edition 2003

For information about permission to reproduce selections from this book, write to
Permissions, W. W. Norton & Company, Inc.
500 Fifth Avenue, New York, NY 10110

Manufacturing by Maple-Vail Book Manufacturing Group
Production manager: Julia Druskin

Library of Congress Cataloging-in-Publication Data

Stott, Rebecca.
Darwin and the barnacle : the story of one tiny creature and history's most
spectacular scientific breakthrough / Rebecca Stott.
p. cm.
Includes bibliographical references and index.
ISBN 0-393-05745-3 (hardcover)
1. Natural selection. 2. Darwin, Charles, 1809–1882. 3. Barnacles.
I. Title.

QH375.S76
576.8'2—dc21   2003002215

W. W. Norton & Company, Inc., 500 Fifth Avenue, New York, N.Y. 10110
www.wwnorton.com

W. W. Norton & Company Ltd., Castle House,
75/76 Wells Street, London W1T 3QT

1 2 3 4 5 6 7 8 9 0

Darwin and the Barnacle

*

Contents

*

List of Illustrations

✳

Acknowledgements

*

Books and ideas grow in conversation and correspondence. The majority of such conversations took place for this book around a large oak table in a seminar room in Darwin College, Cambridge. Here a collection of postgraduates, researchers, writers and academics from widely diverse subjects and disciplines met every fortnight or so for a year to discuss the MS of *Darwin and the Barnacle* as it was being written. To these people I am especially grateful. They include many people from the vibrant History and Philosophy of Science Department of Cambridge University: Jim Secord, Anne Secord, Jim Moore, Patricia Fara, Jim Endersby, Sujit Sivasundaram and Steve Ruskin; from the Cambridge English Faculty, Liz Hodge and David Clifford; from the Cambridge Faculty of Divinity, Thomas Dixon; from the Darwin Correspondence Project, Paul White and Shelley Innes, and Adrian Friday from the Cambridge Zoology Museum.

The book would not have been written without the resources provided by the Cambridge University Library and by the Darwin Correspondence Project. I owe a debt of gratitude for the generosity and support of the distinguished Darwin scholars James Moore, Adrian Desmond and Randal Keynes.

Few historians of science have had the patience or zoological understanding to have made a specialism of Darwin's barnacle work. I am indebted to those who have, particularly Marsha Richmond Shelley White, Phillip Sloan, Alan Love, Bill Newman, D. J. Crisp and A. J. Southward.

Adrian Friday and Prof. Michael Akam gave their time and attention to capturing the remains of Mr Arthrobalanus' dissected body on film. Simon Carnell turned that same body into poetry to form the exquisite epigraph to this book.

At Faber, Julian Loose, Nick Lowndes and Angus Cargill provided intelligent, astute and tough editorial advice as well as meticulous care and attention to detail.

Faith Evans, my literary agent, continues to be a constant inspiration.

Three talented undergraduates from the APU English Department helped in the final editorial stages of the book: Sian Mansfield, Catherine Grant and Alison Smith.

I am also grateful to other richly perceptive readers who have helped shape the book in different ways: special thanks to the remarkable biographer Sally Cline and also to Paul Morrish, James Burgess, Tory Young, Jonathan Burt and Mark Currie. And finally to my son, Jacob, who observed, after I had described the book in the making, that I seemed to have a special fascination with making big things out of little things. So, I explained, did Darwin.

Mr Arthrobalanus

SIMON CARNELL

'Mr' Arthrobalanus
you're an ill-formed monster

with eyes in your stomach, binocular eyes.
Distinctly deviant,

you're an angel
dancing on the pinhead of yourself,

fishing for food with your feet.
Peered hard and long enough at,

you're a grain containing
a strange new world.

From your squatter's nutshell
the whole of creation

is unfolding, like a dropped into water
Japanese paper flower.

Preface

*

The rocks between tide and tide were submarine gardens of a beauty that seemed often to be fabulous, and was positively delusive, since, if we delicately lifted the weed-curtains of a windless pool, though we might for a moment see its sides and floor paven with living blossoms, ivory-white, rosy-red, orange and amethyst, yet all that panoply would melt away, furled into the hollow rock, if we so much as dropped a pebble in to disturb the magic dream.

Edmund Gosse, *Father and Son*

In the summer, my brothers, sister and I lived on the beach at the bottom of our terraced street, foraging, poking about in rock pools, making sandcastles or stone dams, levering shells off rocks to find sea creatures: creatures for dreams and for nightmares. We were always barefoot and the soles of our feet soon hardened to the sharp impress of the acorn barnacle shells that covered every surface below the tide line, millions of white cones – little volcanoes we called them – all different sizes growing over and under and on each other, competing for space, on rocks, on the wooden breakwater, on shells and driftwood. Five heads pressed together, ten feet waving in the air, our bellies on the warm sand, we'd drop a rock or a shell covered in the white cones into a bucket of sea water and wait. First, a tiny hatch opened at the top of each cone. Then a few seconds later, long, feather-like fans unfurled through the hole and began snatching at the water, rhythmically, like a pulse or a heart beat, all together. If we could have made the cone house invisible we would have seen its bizarre inhabitant, a cream-coloured shrimp-like creature, upside down, glued to the rock by its head, fishing for plankton through the hole in its cone with its feathery feet.

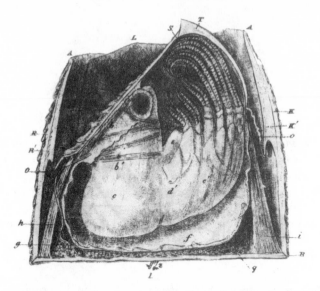

1 A Cross-section of the Acorn Barnacle

Barnacles take two principal shapes: the coned seashore barnacles and the stalked barnacles that cluster on driftwood. As a child I saw the stalked kind for the first time behind glass, not in an aquarium or a museum but in my grandfather's wholesale food warehouse. His father had been a ship's chandler on the east coast of Scotland, but the family business had moved south and was now supplying food to the restaurants and hotels of the south of England. The dimly lit, labyrinthine warehouse was another country: the smells of spices and oils, mountains of sugar and flour sacks, caves of bottled and tinned treasures, trapdoors, levers, pulleys and winches, a locked cold-room that poured out mist and in which hung the smoked carcasses of pigs.

We played hide-and-seek here. One day, seeking a new hiding place, steeling myself against imagined ghosts, I took the little back stairs that creaked up through a trapdoor into a dusty sunlit attic where the boxes of exotic foreign food were kept in jars in stacked boxes: bottled snails, frogs' legs, okra and caviar. That day there was a case I hadn't seen before. It was marked *Perceves*: twelve catering

jars of pickled stalked barnacles from Portugal in a box. They
looked like clawed fingers on the other side of the glass jar, one-inch
prehistoric monsters with stalks the colour of glistening black elephant skins and a claw like the beak of a bird. I couldn't imagine
how anyone could want to eat such things or *how* they would do so.
Snails were bad enough. It made my flesh creep. Why had God
made such creatures; what were they for?

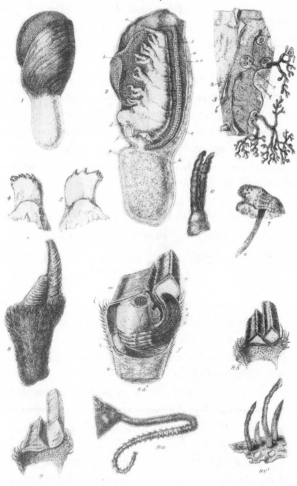

2 A Cross-section of the Stalked Barnacle (*Anelasma: Ibla*)

Thirty years later, a Portuguese waitress in a seafood restaurant overlooking a harbour wall in Viana do Castelo showed me how to deftly twist barnacle stalk from claw and pull out the slither of pink flesh hidden inside the black stalk. It tasted of the sea – mysterious, briny and a little gritty, like mussel flesh. I opened up the beak-like end on my plate, to see the little black creature inside, curled upside down, its feathered feet retracted. In the sea it would be dancing like its coned cousins, unfurled amongst plankton. Now my questions were different: how had it come to this strange shape? How had it *come to be* through unimaginable centuries of metamorphosis? How had it evolved?

By the time I ordered barnacles from the menu in Viana do Castelo, the bizarre creatures had become inseparable in my mind from Charles Darwin, for the great man, the author of one of the most groundbreaking books of all time, had also spent eight years collecting, dissecting, analysing and mapping barnacles. It was a passion – an obsession. It nearly killed him. But the final books, meticulously detailed, won him the Royal Society Medal in 1854 and established him as a scientist who had won his spurs. Without his barnacle spurs and barnacle contacts, *On the Origin of Species by Natural Selection* would have been very differently received.

The barnacle obsession which dominated all Darwin's waking hours between 1846 and 1854 began with a discovery made on the *Beagle* voyage.

On a hot Chilean beach in January 1835 the twenty-four year-old Darwin picked up a conch shell covered in tiny holes, like lace, which he slipped into his pocket. Under the microscope in his *Beagle* cabin, with one of the holes illuminated by candlelight, he saw the squatter, the tiny cream-coloured curled creature that had dug the hole. What manner of beast was it? It looked for all the world like a barnacle, but Darwin knew that, according to the zoology text books, barnacles secrete their shelly homes; they don't *dig* them. He teased out the creature from the base of the hole with a pin and then under the microscope examined its beautiful and complex anatomy. He checked the zoology books in the ship's library just to be sure. It was indeed a barnacle anatomically, although there were unaccountable deviations from the barnacle's archetypal norms. He didn't know it at this point, but this barnacle, soon to be nicknamed Mr Arthrobalanus, would not be finished with him for a further twenty years. This was an encounter on a beach with a creature too small to see with the naked eye that would lead to eight years of meticulous dissection and observation of every known barnacle – fossil and living – in the world and to four published volumes with hundreds of pages of analysis.

Darwin was lucky. The barnacle he had found on the South American beach, with no cone-house or stalk, was an extremely rare form of burrowing barnacle. This squatter was highly unusual, an aberrant, because up until 1830 barnacles had been defined by the shape of their shell-houses not their soft bodies, and this one had no house of its own. For Darwin, anomalies like these raised all sorts of questions about the classification systems themselves. What makes a barnacle a barnacle? And when there is so much variation within a group like the barnacle – in terms of size, method of reproduction and life cycle – what common features hold the family together? His Chilean anomaly would help to explain barnacle evolution and adaptation.

Barnacles had colonised the shorelines, ships' hulls and seabeds of the temperate world. They were ubiquitous. Yet they were as complex and unmapped as the Amazonian rainforest. No one had mapped the barnacle. Darwin would be the first to do so. It would be an evolutionary classification showing how, from a common ancestor, hundreds of different and spectacular barnacle adaptations had taken place over millions of years.

Darwin carried 1,529 species bottled in wine spirits back on the *Beagle* to London in 1836. Amongst these was a single bottle containing a dozen or so very rare minute South American barnacles, teased out from a conch shell and labelled *Balanidae*. He didn't understand them yet, but he would come back to them later. The barnacle was unfinished business.

By 1844, when he was thirty-five, Darwin had formulated and sketched out his species theory in essay form – the incendiary ideas that would change human understanding of time and nature for ever. He had sealed this essay in an envelope, locked it away in a drawer in his study and put together a set of instructions to his wife on how to handle its publication in the event of his death. He now needed to flesh out the theory with evidence and careful rhetoric but, instead of doing this, he turned to the barnacle in the bottle on his study shelf – the riddle that needed to be unlocked. It would take him a month or so to solve it, he thought, and *then* he would complete and publish the species theory.

The baffling, 'illformed' creature he had found in the conch shell was the size of a pin, he remembered. It fitted barnacle body plans in some ways but was completely deviant in others. The more Darwin studied this creature, the more bizarre it appeared to be. How had it come to be this way? How did it fit into the barnacle order? Who were its near relatives?

Within days, Darwin realized that there was no way he could understand just how divergent his Chilean barnacle was until he had seen and mapped most if not all of the hundreds of varieties of barnacles clustered on rock pools, seabeds, driftwood and whale flanks around the world. He began to send for specimens, and once they began to arrive there was no going back; he was hooked. Two years later, his study piled high with barnacle specimens of all shapes and sizes labelled in pillboxes, he had committed himself to writing a definitive monograph. Slowly, he clawed his way back to the problem of the Chilean barnacle, now named Mr Arthrobalanus, via the microscopic examination of all the fossil and living barnacles in the world, nearly twenty years after the first encounter on the South American beach. From 1846 when he began, it took Darwin eight years to work out the riddle of Mr Arthrobalanus's bizarre anatomy, years in which the species theory lay sealed in a drawer in his study – postponed. He was frustrated, confounded, but he couldn't stop.

The story of Darwin's barnacle work is also the story of how scientific discovery sometimes proceeds through indirection, for Darwin's eight-year voyage into barnacle darkness – the last voyage he undertook before publishing *On the Origin of Species by Natural Selection* in 1859 – was a voyage driven by curiosity and obsession but also by an instinct for postponement. Darwin was no doubt hesitating by taking on the barnacles before he went to print with his bold species theory, but it was a particular kind of hesitation, not driven by fear, uncertainty or ambivalence, but by his realisation that the time was not yet ripe for publication of his theory and that he would need to prove himself as a systematizer if he was to be listened to when he did publish.

This passion to classify a commonplace sea creature was not as bizarre as it seems. In the 1830s and 40s invertebrates, particularly marine invertebrates, were at the centre of controversies in taxonomy, comparative anatomy and evolutionary speculation. If all species had evolved from common stocks as some believed, the earliest life forms would have been aquatic single-celled organisms. Thus, Jean-Baptiste Lamarck, the French evolutionist, had claimed in 1809 that invertebrates were the key to understanding how all higher forms had evolved. But the amount of variation within the invertebrate groups made them extremely difficult to classify. As the century progressed, however, and as advanced dredging techniques brought up more previously unclassified creatures from the seabed, microscopes and developing dissection equipment and techniques made it possible for naturalists to see sharper detail inside the bodies of preserved invertebrate specimens, and to record and compare their spectacular anatomies, lifecycles and modes of reproduction.

By the 1840s many naturalists, armed with increasingly powerful microscopes, had taken on particular marine invertebrate groups as classification challenges and opportunities to speculate on a range of conceptual problems in natural philosophy. T.H. Huxley, for instance, ship's naturalist on the *Rattlesnake*, was working on crayfish, squid and jellyfish. Edward Forbes, lecturer in botany at King's College, had finished his book on starfish and had moved on to 'naked eyed' medusae. In the mid 1840s marine invertebrates became critical to debates about sex when a Scandinavian naturalist called Japetus Steenstrup published a controversial study of the reproductive modes of marine invertebrates, showing how extensive multiple-generational asexual reproduction was in nature, particularly under the sea. Sexual reproduction no longer appeared to be nature's dominant method, now that sea creatures were recorded as reproducing every which way: splitting, budding and self-fertilising.

Barnacles were especially challenging to marine naturalists at this time. In the 1830s an army surgeon called John Vaughan Thompson had showed that adult barnacles develop from free-swimming young and consequently were most like crustacea, not molluscs which they had previously been thought to be. Barnacles were now officially misplaced and misunderstood, their place in nature undetermined. But by 1846 when Darwin took out his Chilean barnacle once again, the barnacles were still unclaimed as a classification project. No naturalist had taken on the whole group. No one knew – yet – how difficult the task would be.

Barnacles may have driven Darwin mad, given him nightmares, wrecked his health and his patience, but they were the perfect problem. Barnacles are small. Darwin was astute enough to know this had considerable advantages for him as a systematist. Specimens would travel around the world on mail carriages and trains in small glass jars or in pill boxes. They would *come to him*, posted by an army of collectors, friends, missionaries, friends, naturalists, mineralogists and shell collectors, to his house in Kent. Barnacles cluster on shorelines, they are ubiquitous and easy to collect and preserve. When he had a full collection of specimens, they would be small enough to lay out on a large table in his study so that he could map their diversification from one of the ancient fossil specimens, moving the specimens around to make the continuously branching patterns of a family tree that told an evolutionary story. The perfect puzzle for a speculative man.

The barnacle project would bring Darwin into intimate global correspondence with a community of naturalists, comparative anatomists and zoologists working on similar philosophical problems. He was not a lone genius working in isolation. Through letters and specimen exchanges delivered by an increasingly sophisticated postal system, he worked through his classification and natural philosophical problems in relentless dialogue with hundreds of correspondents, forging ever more nuanced questions and propositions to be answered. In return he was cross-questioned, consulted, challenged and congratulated. By the time he had finished the barnacles, Darwin was at the centre of an intricate web. Each of the thousands of letters he wrote in the barnacle years created another delicate skein in that web, made possible by postal and railway networks stretching out from Kent to numerous European cities and beyond that to Australia, America, and Africa. When he later came to write *On the Origin of Species by Natural Selection* the existence of that web helped to determine the reception of his spectacular and controversial idea. Even those who denounced his theory could not dismiss Darwin as a mere speculator. He was a man who had classified the barnacles, won his spurs, been awarded the Royal Society medal. He was a man of authority and a man with important contacts and supporters.

This book is a reconstruction of Darwin's barnacle voyage pieced together using the thousands of letters he wrote during those years, the books he read, the philosophical questions he formulated for himself and others, and the conversations he had with other zoologists working on similar problems.

Between 1846 and 1854, whilst Darwin remained glued to his microscope, mapping barnacle body parts, all around him the world was changing: his own family swelled and grew from four children to eight; one child – his favourite, Annie – died from a mysterious illness; and beyond his own fine house in Kent the political map of Europe shifted and mutated through waves of revolution. Another intellectual and philosophical revolution was waiting in the wings during these years, already formulated in embryonic form as an essay in Darwin's study drawers: the publication of the *On the Origin of Species by Natural Selection*. But before he turned his mind to the species book, Darwin had to finish his barnacles.

Darwin's encounter with the barnacle began on a Chilean beach in 1835 but his fascination with marine zoology began in Edinburgh in 1825, a fascination shaped and nurtured by conversations with a Scottish doctor and sea sponge expert, Robert Grant, who taught the sixteen-year-old Darwin to dissect sea creatures on the shoreline of the Firth of Forth. He captured Darwin's young imagination, planting philosophical questions about sea creatures and the origins of life that would germinate much later.

1

The Sponge Doctor

*

Were this world an endless plain, and by sailing eastward we could for ever reach new distances, and discover sights more sweet and strange than any Cyclades or Islands of King Solomon, then there were promise in the voyage. But in pursuit of those far mysteries we dream of, or in tormented chase of that demon phantom that, some time or other, swims before all human hearts; while chasing such over this round globe, they either lead us on in barren mazes or midway leave us whelmed.

Herman Melville, *Moby Dick* (1851)

January 1822. Leith, the harbour town of Edinburgh. The weather is so mild that lupins and sweet williams bloom in Edinburgh gardens. From the elegant buildings of old Edinburgh up on the hill crowds of city dwellers follow a black cart carrying two prisoners through open fields down the elegant old Roman road, Leith Walk, to the sands of the harbour town, where gallows have been erected at low tide at the foot of Constitution Street. The cart carries Peter Heaman, from Sweden, and Francois Gautiez, a Frenchman, who had killed their ship's captain and escaped with the cargo of Spanish gold, only to be caught by excise officers in the act of sharing out their gold coins with pint pots on a remote Scottish island. They have been sentenced to death by the High Court of the Admiralty. At the back of the cart, Heaman bows and performs to the crowd, but Gautiez sits silently, hands folded. As the cart turns on to the sands, he looks up to see for a moment the audience that has gathered on the seashore in the January sunshine to watch him die.

Here, on the seashore, crowds from the city join workers from the soap, glass, candle and sailmaking industries of Leith itself, like three great rivers converging. From the west, along the coastal paths, stroll whalers and fishermen and women from the neighbouring fishing village of Newhaven, the women dressed in their distinctive bonnets and striped costumes. From the east stride fishermen from Portobello and those who work in the salt and mining industries of Prestonpans. Children pull mussels off the black rocks underfoot. Seagulls scream. Out in the Firth of Forth, the oyster men fill their boats with paying tourists to provide the best view of all, a ringside seat from the ocean itself.[1]

Just before midday the church choir begins to sing the Fifty-First Psalm, voices struggling against the sound of wind and tide:

> Purge me with hyssop, and I shall be clean
> Wash me, and I shall be whiter than snow.
> ... Then shalt thou be pleased with the sacrifices of righteousness,
> With burnt offering and whole burnt offering
> Then shall they offer bullocks upon thine altar.

The two men shake hands and step forward on to the wooden platform. They have rehearsed this step in their dreams. Arms tied behind their backs they fix their gaze upon the horizon, beyond the grimacing, jeering faces of the crowd, whilst the executioner, Thomas Williams, places a rope around each of their necks and black cloth bags over their heads. The light on the sea is still visible through the thickly woven black cloth as they listen for the sound of the lever moving inexorably into its wooden groove to release the platform beneath their feet. The floor drops away. Bodies spin and jerk in the air like deranged puppets hoisted above their wooden stage. The crowd cheers, throwing paper balls at the twisting bodies like confetti. Startled seagulls scream in the wake of the oyster boats.

Later that afternoon, when the crowds have dispersed and oyster catchers have returned to the winter beach, two porters from the Medical Faculty of the University will cut down the bodies and place them back on the cart. There will be no crowds this time to watch the cart carry the bodies back up the hill, nor to watch the porters carry the corpses through the doors of the Surgeons' Hall and on to the black marble tables of the dissection rooms. Most of the bodies of criminals hanged in Scotland are brought here to the

Medical Faculty housed in the elegant and leafy Surgeons' Square. Corpses are few and far between for Dr Munro and his medical students – a double public execution a rare event. In his dissection theatre, Dr Munro, dressed in a bloodied white apron, and waving a scalpel, will make these pickled bodies last for several weeks, lecturing to a steeply tiered audience of students whilst a paid demonstrator teases out delicate nerves and sinews or saws through limbs.

With cadavers in high demand and in short supply, Edinburgh and its medical schools have become the centre of bodysnatching. Town councils struggle to find funds to build watchtowers and high walls with railings around local graveyards to deter bodysnatchers, but market forces prevail. Bodies fetch money. In Edinburgh, Merryless, Spune and Mowatt, a well-known group of body-buyers, barter openly for the bodies of the recently deceased at the doors of tenement blocks.[2]

Surgeons' Square, Edinburgh. Midnight. Medical school porters patrol the Old Surgeons' Hall, where the stiffening bodies of the pirates lie in an ante-room of the dissection theatre. Otherwise, the square is empty apart from a stray cat or two hunting in the bushes around the lawns; but in an upper window in Dr Barclay's Anatomy School next door, a light flickers. Inside, a well-dressed man in his late twenties, well known to the porters, works alone late into the night, surrounded by human body parts arranged on tables and notes and drawings scattered around the room. The air is thick with the smell of alcohol preservative. Robert Grant is wiry and slight in build, but even at this time of night he is neatly dressed and clean-shaven. He has a high forehead and, despite his youth, he is already balding. He prefers to work at night and, when the anatomy rooms are empty, he often works all night. It is always the way with him. The University porters tidy the room around him before the morning's classes begin. Tomorrow, after his sleepless night, he will join the students to watch the dissection of the pirates.

There is no light in the dissecting room apart from a circle of candles arranged around a watch glass. The lens of Grant's microscope is focused on the watch glass. Inside, lying inert in seawater on the bottom of the glass, there is a small brightly coloured

3 Robert Grant

organism, covered with holes. It is a sea sponge – *Spongia compressa* – retrieved today from the dredging nets of the Newhaven fishermen and passed on to Grant. The fishermen are careful to keep the brightly coloured but inedible sea sponges for the Doctor, to tease them out of their nets and throw them into a bucket of seawater. Dr Grant pays well – more than the farmers who buy barrel-loads of discarded fish parts to use as manure; but the Doctor is particular about the body parts he will buy.

Now Grant peers closely into his watch glass, his body taut and still, for twenty minutes at a time, breaking only to stretch his back and blink his eyes before he returns to the watch. Long periods of time pass like this, in which nothing appears to be moving inside the glass jar or outside in the room itself, except the shadows thrown on the wall by flickering candles.

Tall, clever and cynical, with a sharp tongue and a dry sense of humour, Robert Grant was passionate about sea sponges. The seventh of fourteen children, he had grown up in Edinburgh on the shores of the Firth of Forth, one of the richest marine habitats of the world, especially in sea sponges. He had spent much of his childhood with his parents or private tutor on the beach at Leith, watching the horse races on the sands, or sketching the whaling boats moored in Leith harbour or the oyster boats at Newhaven. Leith races were a particular pleasure: horses galloping the sand racecourse of two miles, Punch and Judy shows, swing-boats, clowns and fortune-tellers.[3] Robert was the child who collected seashells while his brothers wrestled on the sand. He carried them home to arrange them on his mantelpiece amongst his favourite geometry and Greek schoolbooks.

Whilst his brothers went into the army and into the East India Company, Robert signed up for the literature classes of Edinburgh University in 1808, when he was just fifteen; but it was the medical and anatomy classes that held his attention, and soon he had determined to train as a doctor, compelled by the opportunity to study anatomy rather than by any particular desire to heal the sick. The Medical Faculty gave him training in dissection and surgery, but cadavers were scarce; so in order to develop his dissection skills further he joined the Infirmary of Edinburgh as a pupil at the age of seventeen. He signed up as a member of several student societies, where he read research papers and argued with other ambitious young men about the philosophy of human and animal anatomy, the design and structure of bodies and the age of the Earth.

These were exciting times. Fossil and geological discoveries gathered in caves and quarries around the world, and the remains of oyster beds found at the tops of mountains were forcing geologists and comparative anatomists to speculate on the age of the Earth and how it had come to be. This was an age of speculation. There were extraordinary theories abroad that were challenging religious scholarship, and the sea seemed to hold the answer to so many questions. If the Earth had started out as a ball of water, as some argued, the first life forms would have been aquatic. If, as a few others supposed, species had evolved, mutated, over millions of years of unimaginable time, these primitive aquatic creatures were ancestors. Evolutionary riddles were to be pursued in the dark crevices of rock pools and on the seabed.

As President of two reputable student societies,[4] Grant attracted the patronage of the professors of science from his first enrolments in classes. One of these was Robert Jameson, Professor of Natural History, an expert in comparative anatomy and geology, a natural philosopher and one of the most influential men in the University. In 1813 the famous fifty-two-year-old Jameson was formulating a new essay by the French comparative anatomist, Baron Georges Cuvier, *An Essay on the Theory of the Earth*. Like Cuvier and other palaeontologists, Professor Jameson was formulating a theory about the processes that had shaped the Earth since its beginnings. There were many different theories being contested in the universities of Europe, but Jameson was a Neptunist, a follower of the ideas of Abraham Gottlob Werner, a German mineralogist who argued that the earth had begun as a ball of water, a universal ocean, and that all that was now solid on the earth had gradually settled like sediment from that primal water. But Jameson was also interested in other ideas and evidence from Germany and from France in particular.

Throughout Europe natural philosophers were using every fragment of evidence they had collected and labelled – fossils, rock samples, information about plant and animal distribution, anatomical knowledge – to speculate and hypothesize about the beginnings of time. If Jameson and others like him were right and the earth had started out as a ball of water, how had life begun? What were the first creatures on the planet? Why were there shells and the bones of enormous sea creatures to be found thousands of miles inland and at the tops of mountain ranges? Professor Jameson took Grant and his other students into his museum filled with shells, fossils and bones and to the hills around Edinburgh to speculate upon their formation. He translated the new French and German ideas, summarized them in his lectures, and his students debated them noisily in their societies.[5]

By his late teens Grant's studies were directed almost exclusively to comparative anatomy. Like other men of science in Europe he was searching for common patterns in the bodies of animals, patterns of similarity between very diverse species, because these would reveal the essential laws of nature and the origins of life. In preparing for his dissertation in 1813 on the circulation of blood in the foetus, Grant discovered a new book in the University library that opened his mind to new possibilities about the origins of living forms. It was called *Zoonomia, or the Laws of Organic Life* (1794–6) and its author was Erasmus Darwin, an English doctor, poet, inventor and naturalist. By the time Robert Grant read this book in 1813, Erasmus Darwin was dead but his son, Robert, had several children with burgeoning scientific interests including a four-year-old boy, Charles Darwin, already an avid collector of pebbles and plants.

Zoonomia came in two volumes, ran to 586 pages and weighed four pounds. But of all those pages it was the fifty-five-page chapter on generation that excited Grant the most, for in it Erasmus Darwin argued that every living organism on the Earth had descended from *one* common ancestor, a microscopic aquatic filament, and that from this beginning of life all other forms had transmuted and diverged to form all the variety of forms now present on the Earth: 'In some this filament in its advance to maturity has acquired hands and fingers, with a fine sense of touch, as in mankind. In others it has acquired claws or talons, as in tygers and eagles. In others, toes with an intervening web, or membrane, as in seals and geese.'[6] Erasmus Darwin, who had postponed publishing his ideas about the development of species for twenty years, fearful of the consequences, was tentative in his conclusions, even hesitant, carefully honing his ideas into rhetorical questions:

Would it be too bold to imagine, that in the great length of time since the earth began to exist, perhaps millions of ages before the commencement of the history of mankind, would it be too bold to imagine, that all warm-blooded animals had arisen from one living filament, which THE GREAT FIRST CAUSE endowed with animality, with the power of acquiring new parts, attended with new propensities, directed by irritations, sensations, volitions, and associations; and thus possessing the faculty of continuing to improve by its own inherent activity, and of delivering down these improvements by generation to its posterity, world without end![7]

The nature that Erasmus Darwin described was entirely self-sufficient, self-regulating and self-generating; all of its changes could be explained without reference to God and it was still changing, mutating, improving. The word – the dangerous word – was 'transmutation'. The world was still in its infancy, he argued, and the patterns of structure that Grant had seen in animal bodies were, according to Erasmus, the result of common parentage; but when Grant enthused about the brilliance of *Zoonomia* to fellow students and teachers, they either dismissed the book as mere speculation or warned him against the blasphemous ideas that it contained. It was a book by a poet for poets, others said – those with a vivid imagination. Who else could believe that man had grown from primitive aquatic filaments? The man was an inventor of ideas not a researcher, nor an anatomist, nor a dissector. Where was his evidence?[8]

Grant graduated as a doctor in 1814. His father, a solicitor, had died in 1808 when Grant was fifteen years old, unusually dividing up the family fortune in his will equally between the surviving children. Grant was now financially independent, but his money would not finance intellectual independence for more than a few years; so his professors encouraged him to go abroad – Germany and France were the places to go with the questions about anatomical patterns that pressed him. Paris was the centre of comparative anatomy and in 1815, at the end of the Napoleonic Wars, the year after Grant graduated, the great city was accessible again. The seventh son to leave home, Grant's departure caused few worries to his family. His health was good, his professors encouraging, family funds available.

Grant arrived in Paris at the beginning of the winter of 1815–16 when he was twenty-two. The war was over but the city, with its wide boulevards and green parks, was still full of men in uniform: Prussian victors swaggered about the city, boastful and ebullient, filling the nightclubs and dance halls; the British soldiers were more sober and disciplined; even Scottish soldiers in kilts were to be seen on street corners buying lemonade from the street sellers. Grant was one of only a few British medical students in the city, but they lodged together and met in cafés, comparing notes, gossiping about the charismatic and argumentative French professors.

There were many more corpses for dissection in Paris than in Edinburgh. Reforms to medical practice, research and hospital management in the wake of the French Revolution had made Paris the heart of the new medicine, characterized by systematic close observation of human anatomy. The vast Parisian hospitals, no longer in the hands of the Church but in the hands of the republic, had become places of research and teaching and the new generation of doctors was being trained to spend hours sawing, cutting, looking, describing the insides of bodies in order to determine the structure and patterns of disease, only visible beneath the skin. Thirty thousand people a year were treated in the hospitals of Paris and of those who died four fifths were dissected. Body parts on marble slabs, anatomic diagrams on chalk boards, maps of disease, autopsies, microscopes – like a detective, the medical student was trained to watch, listen and record and to trust the evidence of his senses, particularly his eyes, above the authority and abstract, untested theory of old medical textbooks.[9]

Grant spent his first winter studying comparative anatomy in the Musée d'Histoire Naturelle at the Jardin des Plantes, attending lectures by the most famous French comparative anatomists of his day and improving his French so that he could follow their arguments more closely. Carrying letters of recommendation from Professor Jameson, he was immediately admitted into the salons of three French professors: Baron Georges Cuvier, Geoffroy St Hilaire and the elderly Jean-Baptiste Lamarck, Professors of Comparative Anatomy and Zoology at the Jardin des Plantes. Like Grant and Jameson, all three French professors were gathering detailed information about human and animal anatomy. They were trying to understand why there were structural and functional similarities between legs, fins and wings, for instance. Cuvier taught that species were fixed and that any observable similarities between organisms belonging to different zoological families – or *embranchements*, as he called them – were explained by their similar *functions*. Geoffroy, on the other hand, believed in structural relationships that revealed blood relationships across the various groups. He was convinced that there was a *unity of type* and that natural laws could explain the connected body parts of humans and other animals right across the spectrum of creation, but he was not much concerned with transmutation as a way of explaining these common structures.

There was a transmutationist in Paris, however. Lamarck, Professor of Worms and Invertebrates in Paris, was in his seventies in 1815 when Grant arrived and was just completing his seven-volume work *Histoire Naturelle des Animaux sans Vertèbres*. Like Erasmus Darwin before him, he believed that the structural similarities between different animals were due to the development of different species *from a common ancestor*; and that this common ancestor was not a man called Adam, but a primitive marine creature that had appeared through a kind of spontaneous generation. These were controversial ideas because they implicitly questioned the centrality of God in the creation of the Earth, implying that humankind had evolved from a primitive creature and therefore had not been made in God's image. In addition, for these theories to be true, the Earth would have to be hundreds of millions of years old, much older than biblical history allowed. But Lamarck was not setting out to topple the Church or the authority of the Bible – these things were far from his mind. He was looking for a key to nature's riddles and secrets. It was a simple hypothesis: if life started in water and single-celled marine organisms were the first life forms, then, as the 'parents' of all living things, marine invertebrates were the key to understanding the complicated biological processes of all 'higher' animals.[10]

Grant, already seeing nature and transmutation through Erasmus Darwin's eyes, gravitated towards the Lamarck transmutationism, staying in Paris long enough to improve his French, study the contents of the natural history museum and its rich sea-sponge collection of over a hundred different species, and pick up the latest radical philosophical theories shaping zoological interpretation. He read voraciously in Greek, French, German and Italian. The writings of Geoffroy and Lamarck in particular confirmed his belief that marine invertebrates were the key to understanding the origins of life. In this climate of philosophical speculation underpinned by rigorous scientific experiment and observation, Grant determined to find a small corner of the natural world that he could make the subject of his own research, an unknown territory to map. In this way he could work through some of these philosophical problems for himself.

Of all the marine invertebrates discussed in the salons and lecture theatres of Paris, the sea sponge, an aquatic organism full of holes, gave comparative anatomists perhaps the most trouble. The discovery of fossil sponges suggested that this creature was one of the most ancient organisms on the planet, but what kind of creature was it – animal or vegetable? If animal life was to be measured in terms of sensation, movement and the existence of a stomach, then the sea sponge appeared to have few discernible signs of such things – few discernible signs of anything. In fact it was so immobile and apparently insensate that many zoologists until the 1820s had categorized it as a plant. Under a stick a sea mouse would roll, a cuttlefish squirt, a sea anemone retract its tentacles; but prod, poke or pierce a sea sponge and nothing happened. Calm, nerveless and permanently attached to rocks on the seabed, they had been studied by few people. Taken out of the water they quickly died, their prismatic vibrant colours faded. They were an enigma.

The sea sponge was to be Grant's unknown continent. With a clear set of philosophical problems in mind he set off for southern Europe to collect new specimens and to visit as many European marine invertebrate specialists, libraries and natural history collections as he could before his inheritance ran out. These questions took him to Rome, Florence, Pisa, Padua and Pavia, not for paintings or ruins but for books and articles in French, German and Italian – books on marine invertebrates and now sea sponges in particular. He collected and dissected Mediterranean sea sponges on the seashores of Leghorn, Genoa and Venice. He read all the journal articles and books on marine invertebrates in French and German he could find. He took copious notes on the books he read and the creatures he dissected. He asked to speak to marine invertebrate specialists in every city he visited and he found them. He spent eighteen months in Germany, then travelled through Prague, Munich, Switzerland, back down to the south of France and the University of Montpellier. In 1820 he returned to Paris, then visited the natural history and anatomy collections in London, arriving back in Edinburgh by the end of the year with boxes of notebooks, specimens and papers. He was careful with his emerging conclusions, however – careful not to give away his ideas to others.[11]

Grant's actions from 1820, on his return to Edinburgh, are those of an obsessively driven and secretive man, determined to prove a theory, and aware that his time was running out. His inheritance was diminishing and soon he would cease to be a man of science with independent means. Once back in Edinburgh he prepared himself for earning a living as a doctor and as a part-time lecturer in comparative anatomy. He enrolled in Dr Barclay's private anatomy school in 1821, which had the best supply of corpses in Edinburgh, and for two years he was seldom away from the dissection rooms until a friend in 1823 give him the use of his high-walled house on the beach in Prestonpans, a village to the east of Edinburgh, where he could conduct his research in complete privacy.

Dr Barclay's anatomical theories were a long way from the exciting ideas Grant had heard in Paris in conversations with Geoffroy and Lamarck. Barclay, a former clergyman, had set himself against these materialist French ideas. He deplored 'the overweening conceit of the sceptic', and for him those who spoke of 'established laws' were bigots. His own published work was a determined attack on all mechanistic explanations of life. 'Those physiologists who are inclined to favour materialism have never attempted to explain how the first parents of the different species of animals and plants might have been formed,' he complained in 1822.[12] Barclay did not suspect that he had a materialist interloper in his dissecting rooms, or that the beautiful sea sponges stored in Grant's jars were being used as the foundation for a materialist philosophy of nature. He was so impressed with his apprentice's knowledge that in 1824 he asked Grant to give the invertebrate lectures for the comparative anatomy series he had been asked to deliver at the University. During the summer, Grant used the income from these lectures to fund further sponge-collecting expeditions around the coasts and caves of Scotland and Ireland.

The seven sea-sponge essays that Grant finally published in 1825–7 tell a tale of obsession – not the obsession of the collector, determined to add more and more rare specimens to his collection, but that of a natural philosopher. He wanted to understand the complex anatomy of the sea sponge in order to form a cornerstone for a daring philosophy of nature, based on the premise that the sea sponge was an almost unchanged descendant of the very first living forms on Earth. The sea sponge would be the first proof of a philosophy of transmutation, providing a lineage from which higher forms had emerged. So he dissected sea sponges by night, winter night after winter night, for almost the entire five-year period between his return from Europe and his first published sponge essay in 1825. Adult and embryo human body parts by day; adult and larval sponge body parts by night.

Grant's winter residence, Walford House, a large house with a high-walled garden built on the edge of the shore of Prestonpans, was rented from a friend in Dalkeith who used it for his family summer holidays. Prestonpans was established in the twelfth century for the production of salt from water dredged from the Firth and boiled in enormous salt pans. In the eighteenth century, at the Battle of Prestonpans, Highland Jacobite soldiers, carrying scythes and eight-foot-long claymores, massacred hundreds of English soldiers at dawn, mowing them down, and severing limb from limb. The ground was strewn with legs, arms, hands, noses, and mutilated body parts, the ground soaked with gore. By the nineteenth century, however, Prestonpans was a picturesque tourist attraction, offering newly fashionable bathing machines to the summer tourists who rented the large sea-front houses. Surrounded by a high wall and opening on one side through a narrow wooden door, which led down four steep steps directly on to the black rocks and sand, Walford House was the perfect place for Grant to house and dissect his sea sponges through the winter months.

Prestonpans Bay was especially rich in sea sponges. If the Firth was full of sea sponges, the bay, only a matter of feet from Grant's living room, through the garden wall and down the narrow steps on to the rocks, was uniquely structured to gather them, as Grant himself described: 'In Prestonpans Bay, the tide has excavated, in many places, the beds of soft slate-clay from beneath the outgoings of the sand-stone strata, and has thus formed innumerable small caves that are sheltered from the direct force of the waves, by lofty ridges of trap-rocks extending to a great distance from the shore.'[13]

In the spring of 1825 Grant was ready to unveil the results of his five-year investigations. He moved back to Edinburgh. First he decided to test some of the basic premises of his philosophy on the members of the Wernerian Natural History Society, run by Professor Jameson. He turned up to give his very first paper to the Society on 2 April 1825 with several buckets of dead cuttlefish recently dredged from the Firth of Forth: *Loligo sagittata*. Cuvier had argued that no invertebrates had a pancreas; Grant opened up the stomach of one of his Firth cuttlefish and revealed to his audience that what had been thought to be the creature's ovarium was in fact a pancreas. Then he opened another and another, male and female, until the room was filled with the smells and discarded body parts of warm dead cephalopods. There could be no doubt. This proved, he claimed, that the pancreas was to be found much lower in the scale of animals than had previously been believed. There was a pattern to be found running through the entire animal kingdom: humans had pancreases; so had cuttlefish. A few weeks later he was back with buckets of gastropods and sea slugs. These had a pancreas too. Grant was using cuttlefish and sea slugs to call Cuvier's influential map of nature, fixed and divided into absolute *embranchements*, into question.

The Society presentations had been generally uncontroversial, but no one yet realized where Grant's observations about marine invertebrates were leading. He was now prepared to unveil the conclusions he had reached about the ancient, inert, borderline sea sponge. He began to publish his research in serial essay form in 1825 in the journal run by Professor Jameson, the *Edinburgh Philosophical Journal*, which was superseded halfway through the sponge series by the *Edinburgh New Philosophical Journal*. Effectively Grant's seven essays formed a small book, but instead of publishing it in one volume, Grant controlled the stages of his reader's understanding of his work, allowing people time to fully absorb each of the premises from which he would draw his radical conclusions. Each article concluded with the words 'To be continued'.

His very first published essay began: 'Sponges are aquatic productions' – a strange phrase. Stating the obvious, perhaps; but with these words Grant implied, subtly and gently, that the sponge had been produced not by God but by its aquatic environment over eons of time. He was careful with his words, taking his time, careful to imply philosophical conclusions but not to state them bluntly. He built up his seven-part philosophy slowly, layer by layer, premise by premise, taking the reader deep inside the body of the sea sponge with all its mysterious holes and catacombs, to reveal its secrets.

During his years in the libraries of Europe, Grant had read the works of all the eighteenth- and nineteenth-century marine zoologists who had specialized in sea sponges or polyps. His essays are filled with their lyrical European names: Zeller, Lamouroux, Gmelin, Peyssonel, Ellis, Montagu, Pallas, Guettard, Jussieu, Blumenbach, Lichtenstein, Schweigger and Marsigli; but he also returned to ancient knowledge and read Aristotle's work on sea sponges in the original Greek. Aristotle, the natural historian of ancient Greece, was also a specialist in sea sponges, recording, for instance, that the helmet of Achilles had been lined with dried sea sponge. Grant discovered, in returning to this ancient work, that Aristotle, dissecting sea sponges without a microscope on the shores of the Adriatic in 300 BC, had been closer to these creatures than the eminent nineteenth-century professors. He – Robert Grant – would walk in Aristotle's footsteps, not on the shores of Lycia but on the windblown shores of the east of Scotland: 'It is pleasing to observe, that our forefathers, at such a remote period, were occupied, like ourselves, among the rocks of the sea shore, experimenting on this humble and apparently insignificant being.'[14]

Like Aristotle, Grant had come to understand the inner workings of sponges by patient observation and dissection. Like Aristotle, he wanted to know whether they were animals or vegetables or both; how its body worked and how the laws that governed its digestion and reproduction. No one had looked hard enough, watched for long enough – no one, that is, since Aristotle, Grant wrote:

But the philosophy of the sponge, the immutable foundations on which scientific discriminations of the species ought to rest, the minute investigation of the mechanism, the composition, and the uses of all the parts of the animal, and of the extraordinary phenomenon it exhibits in the living state,

– its mode of growth, – its kind of food, – its habits and diseases, – the means of cultivating an animal, which has so long rendered important services to mankind, – its mode of propagating the species, and extending them over the globe, and the great purposes which it is destined to fulfil in the universe, have remained where Aristotle left them, or rather, in this branch of study, mankind have gone backward ever since his time.[15]

One of the primary mysteries of the sponge – whether it was an animal or a vegetable – Grant would have to determine for himself, reaching a conclusion where scores of other naturalists had failed to do so. To establish that it was an animal, Grant had to prove three things: that it was sensitive, that it could move independently and that it had a stomach and anus for digestion. His experiments began with an attempt to discover the purpose of the holes that covered the sea sponge's body: were these the elementary forms of a digestive system? Some experts thought that these holes were made by other parasitic creatures burrowing into the soft flesh. Other zoologists believed that sponges were vegetables and therefore argued that the holes were like the pores on the surfaces of leaves, holes through which the vegetable absorbed water. Grant knew that the only way to settle this matter once and for all was to keep watch – to watch the holes of the sponge night and day under a microscope.

Discovering that the single, intense light of a candle enabled him to observed more effectively than the diffuse light of day, he soon developed a habit of entirely nocturnal research at Prestonpans. He made his first major breakthrough some weeks after he began this watch, alone in the middle of the night, with only the sounds of the waves and an occasional lone gull to break the silence:

On moving the watch-glass, so as to bring one of the apertures on the side of the sponge fully into view, I beheld, for the first time, the splendid spectacle of this living fountain vomiting forth, from a circular cavity, an impetuous torrent of liquid matter, and hurling along, in rapid succession, opaque masses, which it strewed everywhere around. The beauty and novelty of such a scene in the animal kingdom, long arrested my attention, but, after twenty-five minutes of constant observation, I was obliged to withdraw my eye from fatigue, without having seen the torrent for one instance change its direction, or diminish, in the slightest degree, the rapidity of its course. I continued to watch the same orifice, at short intervals, for five hours . . . but still the stream rolled on with constant and equal velocity.[16]

'Vomit' hurt and 'strew' are peculiarly violent words for an animal known for its silent, stubborn stillness; however, Grant soon came to see that he was not watching vomiting but excretion. He was the first to observe the excretion of the sea sponge at close range and it was for him a splendid spectacle, because it had been so long awaited and because it showed evidence of digestion. This animal can indeed excrete for five hours continuously through a single hole and, now that he had witnessed this spectacle of excretion with his own eyes, Grant could confidently call this hole a 'fecal orifice'. His revelation occurred in the middle of a November night with no one to share it. He had discovered the anus of the sea sponge – or rather he had discovered that sea sponges are covered with anuses; and where there is an anus there is a stomach. He was putting together incontrovertible evidence that the sponge had animal characteristics.

The next question for Grant was about the strength of the current of the 'fecal fountains'. For several nights he experimented with blocking the anal orifices of the sponges with a variety of domestic objects to determine the strength of the current: pieces of chalk, cork, dry paper, soft bread, unburnt black coal – almost anything to hand. He recorded all the experiments in detail in his notebooks. Only a drop of mercury was heavy enough to block the stream. So Grant concluded that, whilst some of the holes on the body of the sea sponge are 'fecal orifices', the others are 'pores' used for ingesting food, like *mouths*. The holes that lead into the labyrinthine passages of the sponge afford a constant passage of liquids: food passing in and opaque fecal liquids passing out. Aristotle had known this too; this was where being able to read Aristotle's original Greek text was important for Grant, for he could show that the words Aristotle used meant 'pore' and 'orifice'.[17]

Grant, however, was not so sure that it was definitively an animal. He had a hunch that it was an intermediate form, a link between the animal and vegetable kingdoms. For the final proof of its animal status, Grant had to prove that the sponge was sensitive, another important area of dispute amongst zoologists. So Grant set to work; over the nights of three winters he systematically tortured the sea sponges in his collection in Prestonpans in an inventive series of ways, but failed to find any evidence of sensitivity: 'I have plunged portions of the branched and sessile sponges alive into acids, alcohol and ammonia, in order to excite their bodies to some kind of visible contractile motions, but have not produced, by these powerful agents, any more effect upon the living specimens, than upon those which had long been dead.'[18]

The third area of investigation was the question of voluntary movement: animals move independently; plants don't. In his long nocturnal vigils in late autumn, Grant watched a sea sponge excrete eggs – or ova – through its holes, and under his microscope he saw to his amazement that the eggs seemed to be moving by themselves. Under a more powerful lens he was able to see that these eggs were covered in small hairs (*cilia*), which propelled them along, away from the parent sponge. The parent sponges might have been stubbornly inert then, apparently insensate, but the young ones moved like things possessed, gyrating and twisting and wriggling together, propelling themselves away from their parents. Grant was sure there was an evolutionary reason for this: the propagation of the species depended upon the spontaneous motion of the young, their ability to reach new breeding grounds and disperse the species. The enigmatic and peculiar sponge was perfectly adapted, then, to survival in the deep sea – so perfectly that it had not needed to evolve any further, given that its aquatic conditions had remained relatively stable: 'This animal . . . seems eminently calculated for an extensive distribution, from the remarkable simplicity of its structure, and the few elements required for its subsistence.'[19]

By 1826 Grant was claiming that his evidence proved that the sea sponge, which had changed relatively little since prehistoric times, was indeed an animal but one that was so close to the frontier between animal and vegetable kingdoms as to be virtually on it. By comparing sponge ova and those of other simple living organisms with the ova of the algae, he argued that there was a common monadic base for plants and animals. They had similar components. Somewhere in their ancestry there had been a meeting point.[20] Grant had been able to reach these conclusions in part because his European tour had taken him beyond Paris and Lamarck and into Germany where, in the University of Heidelberg, he had met the young Professor of Anatomy and Physiology, Frederick Tiedemann. Lamarck, for all his transmutationism, still believed in an absolute demarcation between animal and vegetable kingdoms. Tiedemann, however, believed that in the most simple and ancient life forms the boundaries between these kingdoms were not fixed. Grant's sea-sponge work had helped him to think through these problems; but he was a long way from establishing a working theory as yet.

Grant's tireless series of seasponge observations, published between 1825 and 1827, formed part of the burden of proof necessary to support a theory of transmutation of species that had so far been held by most as mere speculation. But it was only a beginning. Grant had spent five years dissecting, watching and describing sea sponges, ancient, enigmatic and entirely borderline creatures. Other borderline creatures from the seabed would need to be examined by men and women with the patience, determination and philosophical drive of Grant if the secrets of the aquatic origins of life were to be fully understood. New observers and natural philosophers were needed.

It is 9 February 1826. Charles Darwin, sixteen years old, stands amongst the black rocks on the sands at Leith at the foot of Constitution Street, making notes in a small red notebook, while all around him bare-footed fisherwomen and their children collect mussels to bait their lines. He has enrolled as a medical student at the University, but, bored by the lectures, he escapes to the winter seashore every few days. His boots are salt-stained. The leather-bound notebook is a diary called *The Edinburgh Ladies' and Gentlemen's Pocket Souvenir for 1826*, a present from his elder and much-loved brother Erasmus Alvey Darwin – 'Ras' – also a student at the University. The diary lists tide times alongside a list of the dates of accessions of the Scottish kings. Darwin's handwriting is small and neat despite the fact that his hands, ungloved, are cold and stiff in the biting February sea winds. The paper of his notebook flaps about. The bright scarlet leather is good quality; it has a leather fastening flap. Inside, down the seam along the spine, there is a sheath designed to hold a very small pencil, but Darwin writes in ink and has already marked it as his own, signing and dating his name confidently as 'C. Darwin, Jan 1st, 1826' inside the front cover.

Behind him the bathing machines used by the summer tourists are covered in tarpaulin and stand in a line in front of the warehouses and cone-shaped brick kilns of the Leith Glassworks Company. To his left, beyond the Martello tower built to defend the town against French invasion during the Napoleonic Wars, the masts of moored boats in the harbour look like a flagged winter forest on the horizon. The whaling boats will sail for the Davies Straits or Greenland in March, but in January the men work on the boats or in the boiling houses that back on to the sands next to the glassworks, boiling the blubber to make soap and candles, or packing up the whalebone for the corset makers of Edinburgh. The air is full of the pungent oily smell of blubber.[21]

4 Leith Races

Darwin crouches to peer at a small hairy creature, marooned in one of the pools between the black rocks on the rippled grey sand. It looks like a mouse; in fact the fishermen call it a sea mouse. An incoming wave washes the body clean of the sand that covers it, revealing a green, brown and gold furry body, about three or four inches long. Darwin writes its Latin name neatly in his scarlet notebook: *Aphrodita aculeata* – 'Stinging Aphrodite'. Linnaeus named this creature after the Greek goddess of love, who rose from the waves, but this creature is neither goddess nor mouse. It is a parasitic worm that lives buried in sand on the seabed and is only rarely found washed up on to the shore, usually after a storm.

Darwin reaches for a stick to prod at it. He knows its hairs will sting him painfully. Irritated by Darwin's stick, the mouse recoils and tries to roll itself into a ball. Almost dead, it moves very slowly. Darwin is excited. Although they are not uncommon on the Leith sands, he has probably not seen a live sea mouse before; but he has read about them. He remembers that the marine invertebrate specialists Turton and Linnaeus disagree about the number of feelers it has. He records all these experiments and descriptions in his scarlet diary: 'Caught a sea mouse Aphrodita Aculeata of Linnaeus; length about three or four inches, when its mouth was touched it tried to coil itself in a ball, but it was very inert. Turton states that it has only two feelers, does not Linnaeus say 4? I thought I perceived them.'[22]

Mating wagtails and swallows gathering in the warm winds above Edinburgh will distract his attention from the beach in early summer; but in the spring of 1826 it is sea creatures Charles Darwin wants to know about. So he has come here to the shore between Leith and Portobello to see what the night's storms and spring tides have washed up. He has sand in his shoes. He will come back to this shoreline every few days over the following weeks throughout March and April, walking the fifteen-minute walk from his lodgings down to the seashore past the caravan shows, street singers and organ grinders of Leith Walk. It is here on the Leith seashore that he will meet Dr Grant, the philosophical sponge doctor.

2

Riddles of the Rock Pools

*

Before the land rose out of the ocean, and became *dry* land, chaos reigned; and between high and low water mark, where she is partially disrobed and rising, a sort of chaos reigns still, which only anomalous creatures can inhabit . . .

The sea-shore is a sort of neutral ground, a most advantageous point from which to contemplate this world . . . The waves forever rolling to the land are too far-travelled and untameable to be familiar. Creeping along the endless beach amid the sun-squall and the foam, it occurs to us that we, too, are the product of sea slime.

Henry David Thoreau, from *Cape Cod*

From Carlton Hill, in Edinburgh, Darwin's eyes would have followed the line of the wide street of Leith Walk, lined by fields and terraces of elegant houses, all the way down to the harbour of Leith. Beyond the harbour, the Firth of Forth, grey and steely in the winter light, stretched out from left to right, peppered by fishing boats and single-masted oyster boats. The shoreline view in 1825 was dominated by four cone-shaped brick mounds down on the sands of Leith, like giant brown barnacles between Darwin and the sea, which only a year before would have trailed smoke into the sky. These were the kilns of the Leith Glassworks Company, which had fallen into bankruptcy in 1825 because of high taxes on glass. They were already falling into dis-repair; but the sea-bathing industry was growing. Seafield Baths, with seventeen hot, cold and tepid baths, a large plunge bath and hotel accommodation, had been trading successfully since 1813 and the number of bathing machines lined up on the sands grew from year to year. A regular plunge into a cold sea, doctors claimed, would restore an ailing body to a natural state of health. The sea acted as an antidote to the corporeal corruptions and pollutions of the industrial city, by restoring balance and circulation.

The two brothers, Charles and Ras Darwin, had arrived in Edinburgh in late October 1825. In studying medicine at Edinburgh, Darwin was walking a family route – his brother, father and grandfather had all enrolled as medical students there. After spending a few nights at the Star Hotel in Princes Street, they found lodgings four flights up at 11 Lothian Street with a Mrs Mackay: two bedrooms and a sitting room that were 'very comfortable & near the College'.[1] Darwin's first impressions of Edinburgh were of its gloominess; but their lodgings, he noted in a letter to his sister, were remarkably well lit compared to some of the 'little holes' he had seen. Light would be important here – for reading, for writing up his lecture and reading notes, and for his microscope.

For the next two years these small rooms would be a place of intense concentration, for although Darwin was later critical about his experience of University life in Edinburgh and gave the impression that he didn't achieve much, he and his brother read voraciously. The curriculum at Edinburgh was based entirely upon lectures, some held as early as eight o'clock in the morning; but they were not enforced, and Darwin found them very varied in quality, as he wrote to his sister in January 1826:

As you know nothing either of the Lectures or Lecturers, I will give you a short account of them. – Dr Duncan is so very learned that his wisdom has left no room for his sense, & he lectures, as I have already said, on the Materia Medica, which cannot be translated into any word expressive enough of its stupidity. These few last mornings, however, he has shown signs of improvement & I hope he will 'go on as well as can be expected.' His lectures begin at eight in the morning. – Dr Hope begins at ten o'clock, & I like both him & his lectures very much. (After which Erasmus goes to Mr Lizars on Anatomy, who is a charming lecturer) At 12, the Hospital, after which I attend Munro on Anatomy – I dislike him & his Lectures so much that I cannot speak with decency about them. He is so dirty in person & actions – Thrice a week we have what is called Clinical Lectures, which means lectures on the sick people in the Hospitals – these I like very much. – I said this account should be short, but I am afraid it has been too long like the Lecturers themselves.[2]

Darwin confessed to his sisters in letters that he found all his lectures dull and preferred to follow his own intellectual curiosity in self-directed reading; but his elder sister, Catherine, who had filled his mother's role after his mother's death, was concerned and spoke to their father. Robert Darwin, a powerful man of great direction and drive, was shocked at his son's lack of application: 'If you do not discontinue your present indulgent way, your course of study will be utterly useless,' he wrote fiercely, dictating to Susan.[3] He was also anxious about the health of his two sons in the fierce Edinburgh winters and advised them to remember to wear the special fleece vests he had sent them. Ras, instinctively reclusive and frail in health, was reluctant to socialize with fellow students, so the brothers had a good deal of time on their hands for reading. Between them, they signed out more books from the University library in their first term than most other students did in that year. Among these were books on zoology, at this point in the history of science one of the most philosophical of the sciences.

Although Darwin had been sent to Edinburgh with the expectation that he would become a doctor, he quickly found that surgery turned his stomach and that his intellectual interests ranged beyond medical questions. He also discovered in 1826 that his future inheritance would be large enough to make it unnecessary for him to earn his own living. This belief, as he said himself, 'was sufficient to check any strenuous effort to learn medicine',[4] but it also gave him the freedom to be an intellectual dilettante and in an environment that encouraged independence of mind. He had the butterfly-like curiosity of a sixteen-year-old. In January 1826 he may have been skipping lectures, but he was busy filling his notebooks with zoological observations and taking private taxidermy lessons from a black taxidermist who lodged in the same street.[5] He wanted to be able to stuff birds.

Freed from parental authority, the two brothers roamed the city, drinking porter and eating oysters in the candlelit taverns, visiting the theatre, dining with old friends of their father's. It was a city full of drama and contrasts, with high-rise tenements built on the steep sides of the craggy hills. A sharp and steep ravine dissected the city so that the two sides had to be linked by bridges – a jagged urban skyline surrounded by jagged hills; and a castle perched on top of a mountain in the midst of it all. But whilst Edinburgh may have been elegant – 'stately Edinburgh throned on crags', as Wordsworth called it – the Old Town slums were dangerous and desperate places. The brothers explored the markets held down in the deep ravine that divided the city. Behind the covered stalls of the fish market, rows of fisherwomen bantered loudly with their customers over the wicker baskets of pilchards, cod, haddock, herrings, mackerel and lump-fish, demanding three times the price they expected so as to start the bartering high. Darwin would often have woken to the sound of the cries of fisherwomen selling clams, mussels, prawns and oysters on Edinburgh street corners in the early morning.

Meanwhile, Darwin was using his private income and the marine invertebrates he found at Leith to think through philosophical questions about the origins of life, questions that were now of a distinctly aquatic character. In conversation with the other young men of the student societies of Edinburgh, he would continue to philosophize about the sea creatures he found at Leith and Newhaven. Like Grant before him, he had read his grandfather's book, *Zoonomia*, in 1826, and like Grant he had questions about the body structures of certain marine invertebrates that the shores of the richly fertile Firth of Forth could help him answer. Later, at Cambridge University, with no ready supply of sea creatures to dissect, he would turn his attention to the body structures of insects and begin to see important points of connection between them and certain sea creatures; but here the zoological riddles that taxed his mind came chiefly from the rock pools of the Firth of Forth.

That year Darwin was also reading a book by a notable Scottish naturalist, John Fleming, called *The Philosophy of Zoology*. Fleming, a minister of Flisk, a tiny fishing village with a population of 300 on the banks of the Tay in Fife, was a passionate collector of sea crea-tures, but he rejected transmutation or development as a way of explaining the variety of anatomic structures under the sea. These ideas had not yet been absolutely proved, he claimed. Yet his zoo-logical observations of marine creatures were making him face up to disturbing questions. He couldn't deny that some sea creatures seemed to slip between the absolute categories marked out by

Cuvier. Creatures like the infuriating sea sponge defied established systems of classification. Although there was much to suggest that the animal and vegetable kingdoms had once been joined,[6] he wrote, there were some absolute differences between animals and plants that could not be denied. He was working with a similar set of definitions to Grant's: sensation, independent movement and digestion were the three critical distinctions between animals and plants. These were also Darwin's seashore preoccupations from as early as January 1826. Grant's, too; but the two men had not yet met.

In February 1826 Darwin returned with his brother to the sands of Leith. Ras, wading out in the water, caught a cuttlefish in his dredging net, a weird, squid-shaped cephalopod, about eight inches long, with a ring of ten arms round its large mouth and a body sac filled with blackish fluid. The brothers watched it thrash about in the net, then released it to watch its strange movements in the water. Darwin noted in his red notebook that its mouth 'had a bill, like a parrot's' inside its ring of suckered arms. It had huge eyes, too, bright lizard-like eyes on either side of its beak. It hovered and darted about very quickly in the water like a jellyfish, all tentacles and suckers, fanning its fins and sometimes forcing a stream of water, siphon-like, through a body opening behind its head. Darwin checked its classification back in the textbooks in Lothian Street: *Loligo sagittata* – Barbed fish.

Two days later Darwin explored the ridge of rocks that at low tide look as if they have been scattered along the grey sands from Crammond round to Prestonpans. They were – and still are – black with mussel shells. Against this gleaming black rocky crowd of mussels and barnacles, growing in and over each other in their thousands, Darwin's eyes caught sight of a flash of bright orange, a globular zoophyte. He prised it out from its rock hole and placed it in a basin of seawater – a basin he had carried with him down the hill from Edinburgh. It immediately turned itself inside out like a glove. So he prodded it with his stick and it reversed the process. This time he carried the basin back up Leith Walk to his lodgings and slipped the animal into a jar of wine spirits, where it instantly ceased its aquatic contortions.

In the final days of February the night's storms took Darwin back down to the sands. Here he found that the pools of water left by the sea around the black rocks were full of bright-red sea anemones. Under the still-stormy sky and against the wet, black rocks they gleamed like glossy strawberries. Gingerly, he pulled one or two from the rock and placed them carefully into his pan of seawater. They were cold and slimy, difficult to grasp. Like the zoophyte, once pulled from the safety of the rock and frustrated by the slippery surface of Darwin's bowl, they gyrated and twisted and turned themselves inside out, entirely changing the shape of their bodies. Inside each of the strange cylindrical rocks that covered the black rocks he found a *Pholas candida*, about half an inch deep. These sack-like creatures had bored into the holes by emitting phosphoric acid. At night the rocks, lit up, glowed under the moonlight.

The shore itself, after the night's storms, was littered with the bodies of still-live, twitching cuttlefish; there were glossy tentacles everywhere on the shoreline, twisted up with the dark-green sea weed, mussel shells and driftwood. They were enormously sensitive, Darwin noted. He prodded them with his stick and they squirted a dark-coloured liquid on to the sand. Darwin was sure that they had begun to squirt even before he touched them: they seemed to be sensitive to the slightest movement. So he experimented, creeping up on them to monitor their sensitivity. Everywhere along the sand the cuttlefish were oozing black ink as Darwin played Grandmother's Footsteps with them. This cuttlefish liquid is called sepia, which, when dried to a fine power, was used in ink production. The brown ink that Darwin used in his red notebook in 1826 may even have contained some of this pigment derived from dried cuttlefish sacs. Before he turned away to wander further along the sands, he lifted a few limp cuttlefish back into the water to watch them swim.[7]

Darwin picked up a starfish here, too, with only three arms, the other two having been torn off by the violence of the storm-driven waves; but he noticed in wonder that even just a few hours since the storm had abated, two new arms had begun to grow. The starfish repairs itself again and again. In March, alone in Edinburgh after Erasmus's departure for London, he found spawning starfish and more sea mice at Leith. Growing more confident, he picked them up with his stick and threw them back into the sea, where they rolled themselves up like hedgehogs. The shore in March was full of hideous, jelly-like creatures: a yellowish globular mass called a sea wash ball, a three-foot green worm with numerous feet on each side and a nose an inch long, with dot-like eyes.

So why so much prodding and poking? Why was the sixteen-year-old Darwin so interested in the acrobatics of these creatures – so much so that he would prise them off their rocks in order to watch them dance in his bowls of seawater? What questions was he asking of the cuttlefish and the sea mice in Leith in 1826? The experiments he conducted on these creatures give us further clues about his questions. His notebooks show that he was interested in how they moved and how sensitive they were, because these questions would help him understand their place in nature, particularly their animal or vegetable status. These questions had also troubled Erasmus Darwin and Jean-Baptiste Lamarck, and ten miles east round the coast in a high-walled house in Prestonpans, they continued to trouble Robert Grant.

That summer, away in Wales on a walking holiday during his university vacation, Darwin missed seeing a luminous meteor that lit up the evening skies of Edinburgh on Sunday, 27 August. The sky hung heavy for some time beforehand, meteorologists noted. Suddenly a large body of fire, pear-shaped, and apparently the size of a beehive, moved across the sky from the south-west to the north-east with a rushing noise and turned the night sky as bright as day. It left behind it a very long train, not of sparks, but fluid-like and of the brightest prismatic colours, mirrored in the waters of the Firth of Forth.

John Coldstream, another young medical student at the University and now a protégé of Robert Grant's, was on the shore of Leith to see the comet and record it in his diary that evening. He had a particular interest in comets, the aurora borealis and the sea sponges of the Firth of Forth. Like Grant, he had been born at Leith, attended Edinburgh High School and graduated to the Medical Faculty of the University. Like Grant, he had been building an invertebrate collection since childhood. Unlike Grant, he was religious. However, the more he

5 Leith Pier and Harbour, 1798

collected and studied marine creatures, the more his scientific and religious beliefs came into conflict. In the early days, when he was a member of the Leith Juvenile Bible Society, the abundant life of the shoreline had been a constant testimony to the benevolence of God. At fifteen, he had stood on the sands of Leith at dawn on a winter morning in 1821 and recorded the sight in his diary:

This morning as the storm subsided, I determined to go down to the sands at Leith that I might revel in the riches which might have been cast up by the deep after the terrible storm. First I went to the end of the pier. I saw a most beautiful sight. As the sun was rising, his rays dissipated a thick frosty fog which hovered around; this rose like a curtain on the horizon, and displayed the whole coast of Fife clothed in snow, and reddened by the sunbeam; a large number of vessels lying at anchor in the Roads on account of the storm; dark blue waves rolling in silent majesty, undriven by any wind, and all forming the grandest coup-d'oeil I have ever seen. Returning thence, I strolled along the sands, past the baths, and picked up a number of curious things. There was a scarcity of shells, but asterias, actinae, aphrodites, and crabs were very common. I brought several home, along with a fine specimen of the Cyproea islandica, a new coral &c. I added an Aphrodite aculeata to my menagerie of living animals.[8]

From 1823, when his marine interests brought him into the company of the sponge doctor, Robert Grant, Coldstream's diaries record a sudden change in his personality and outlook. He was seventeen when he met Grant and his journals from 1823 record a painful story of self-blame, torment and physical self-loathing. On his eighteenth birthday, 19 March 1824, for instance, he wrote:

My praise is altogether an unclean thing; my glorifying of the Lord is filth-iness before Him. From the dust do I cry unto thee, O God! Hear me, hear me. I earnestly beseech Thee to purify my heart . . . I am at a time of life when the amusements of this little world lead me away into temptation; now, heavenly Father, point out to me in what measure I should best enjoy these, that all my conduct may be to Thy glory. Oh, were I prepared, how I would fly from the attractions of the flesh.[9]

But Coldstream didn't leave the Plinian Society, nor the company of the doctor. He continued to struggle with his religious doubts and with a certain mysterious sense of personal corruption focused on his own sexual body during these years.[10] He became one of the Society's presidents in 1824–5 and one of the doctor's closest companions, collecting sea sponges for him and for the Museum of Natural History at the University. He also was always on the lookout for other young naturalists who might be nominat-ed for membership of the Society. He published regularly in the *Edinburgh Philosophical Journal* – articles on the springs of Ben Nevis, the saltiness and transparency of the water of the ocean, on hoar frost, on the aurora borealis, which he saw out in the Firth of Forth on his nineteenth birthday, in 1825, and on the sea sponges and the zoophytes of the Firth of Forth. The Firth, its sea crea-tures and meteorological phenomena were the primary source of his scientific and philosophical speculations; but they were also to drive him to ask questions he did not want to ask and which would eventually drive him to moral and religious breakdown in 1828.

Darwin returned to Edinburgh University in the October of 1826, alone. His brother Ras had completed his Edinburgh studies that summer and was now living and studying in London at the Windmill Street Anatomical School. In Ras' absence, Darwin would have to find alternative shore companions with whom to debate the philosophical zoological questions which pressed upon him. In November 1826 Coldstream nominated Charles Darwin for membership of the Plinian Society, but it seems likely that the three men – Grant, Darwin and Coldstream – had already met on the sands of Leith, for Coldstream walked on Leith beach most days with his notebooks and thermometers, taking down detailed measurements of the meteorological conditions for his regular column in the *Edinburgh New Philosophical Journal*, or looking for sea sponges with Robert Grant. Coldstream at this time was Grant's companion and assistant. They might have identified Darwin as a potential acolyte from some distance away, his tailored clothes, collecting equipment and notebooks marking him out from the other occupants of the beach, the fisherwomen and children collecting baskets of mussels on the black rocks to use as bait for fishing lines. Darwin was a fine catch. They would not have known at that first encounter that this athletic, tall, well-dressed young man was the grandson of Erasmus Darwin himself.

The student-run Plinian Natural History Society met every Tuesday evening in an underground room in the University. It had about one hundred and fifty members, but usually there were about twenty-five young men at each meeting. Hungry for new ideas and a forum for debate, Darwin was a regular attender that winter; between his election in November 1826 and his leaving Edinburgh in April 1827 he attended all but one of the nineteen weekly meetings of the Society, a marked contrast to his erratic attendance at lectures. His contributions must have been engaging from the start because only a week after his election as member he was elected to the Society Council.[11] Many of the members were young medical students of Darwin's age, some, like Darwin, still in their teens. Darwin was seventeen, William Kay, from Liverpool, and George Fife, from Newcastle, were nineteen; John Coldstream, from Leith, and William Ainsworth, from Lancashire, were twenty; William Browne, from Stirling, was twenty-two. Several of the members were involved in radical politics. There were five young presidents that winter: Ainsworth, Coldstream, Kay, Browne and Fife. These, with Darwin, appear to have been the core of the group; but there was a much older man who dominated the meetings: Robert Grant, now thirty-four.

There were several short papers given every week by members and invited guests. Many had an emphasis on local natural history or were based on natural-history expeditions undertaken by the members. Students gave papers on methods of obtaining bromine from soap-boilers' waste and on its nature and properties (Leith had a soap-making industry), on the capture of whales on the coast of the Shetlands, and on the oceanic and atmospheric currents. Darwin stated in his *Autobiography* that, unlike the University lectures, he found the Society meetings stimulating. This was a stimulating time; the papers were often radical and controversial. William Greg, also seventeen when he was elected to the Society at the same time as Darwin, immediately offered a paper, for instance, that would prove that 'the lower animals possess every faculty & propensity of the human mind'.[12] Grant's heretical ideas about the lowly origins of life were entirely at home in this company of radical young men. Even if members like Coldstream found such ideas disturbing, they were clearly compulsive and fascinating.

Darwin's friendship with Coldstream blossomed in the winter of 1826–7 in the meetings of the Society, in their winter walks along the Forth with Grant and in the natural history lectures and museum of Professor Jameson. Darwin was alive to the contradictions in Coldstream's character from the start, describing him as 'prim, formal, highly religious and most kindhearted'.[13] Coldstream took him out on the dredging boats that sailed from Newhaven, the centre of the oyster trade in the Firth and only a mile to the west of Leith. In the early nineteenth century the houses of this fishing town that lined the shore were red-tiled, two-storey houses with outside stairs and jutting gables. Oyster boats lined the sands and whaling boats filled the harbour. Oysters were big business in the early nineteenth century. Thirty million oysters a year were dredged up from the Firth of Forth. The fishwives, with their striped dresses and baskets slung over their backs, were already an attraction for artists and tourists from Edinburgh and London.[14]

Given Coldstream's meteorological interests, it is likely that Darwin and Coldstream would have seen the startlingly vivid northern lights that appeared in the night sky on 16 January that winter. David Blackadder, another member of the Edinburgh student natural history societies, reported the phenomenon in the *Edinburgh New Philosophical Journal*:

Evening fine, brilliant moonshine, a beautiful, opaque, drapery of cloud . . . the air was now calm and serene. In the course of a few minutes the cloudy tissue had entirely disappeared. A brilliant aurora was exhibited. In rapid change of feature and distinguished by unusual proximity. Horizontal cloudy vapours, of great tenuity, repeatedly accompanied its more brilliant evolutions, seeming to support its columns; appearing and vanishing with the more vivid coruscations. Thereafter, the aurora became extended, to right and left, forming the segment of a large circle . . . the arch always presented a broken or interrupted line, with recurrence of separate masses of luminous spears, of a brilliant bluish-white lustre; a golden tint, and burnished lustre, distinguished the continuous arch of the central portion, which, towards the left, became coppery.[15]

Even if the winter skies were dazzling, the dredging expeditions would have been cold and cramped for the young naturalists. The oyster boats were small and carried four men, who left at daybreak and rowed and dredged all day, sometimes for twelve hours. They sang fishing shanties and the rhythms of the boat, the movement of the oars, the dropping and hauling in of the dredge all followed the cadences of their singing. By the end of the day their boats would be heavy with gleaming oyster shells, many covered with barnacles. There was danger too. The men of Newhaven were often in territorial disputes with the oyster dredgers of Prestonpans. These attacks took place on the open sea with each boat trying to capture the other. Darwin, who suffered from acute seasickness later on the *Beagle*, cannot have found these dredging voyages easy, despite the pleasures of teasing out sea creatures from the nets and dredges and slipping them into glass jars. However, in the company of friends like Coldstream, Grant and Ainsworth, he came to know the Forth well that winter, travelling by boat as far as the coast of Fife, and sometimes to the islands of May and Inchkeith. On one occasion a storm drove Darwin and Ainsworth to take refuge at the lighthouse on Inchkeith island.[16]

The Plinian Society had its measure of practical jokes as well as philosophical speculation. In the spring of 1827 Darwin and William Kay, then the secretary of the Society, co-authored a spoof paper for the Society called 'A Zoological Walk'. It was a satire on the type of lengthy, inconsequential journal-style paper regularly given by another student called Ritchie. It was written to be delivered in alternate voices by Darwin and Kay. The paper described every minute detail of their long walk from Leith to Portobello during which the shoreline was shrouded in thick fog. When they arrived at Portobello the tide had come in:

Here from the first we had hoped for an ample field of amusement; but alas, again we were deceived, the shore appeared perfectly destitute and void of everything that could interest the zoologist. We then, as a last resource, determined to proceed to some rocks lying a mile or two to the right of Portobello . . . when we gained this our last destiny, our last long looked for scene of amusement, [we found] the few last jutting points covered by the ebbing tide. We lingered a few moments then turning our backs on the hidden treasure, disappointed, we retraced our tedious footsteps to Edinburgh.[17]

On 16 March 1827 Darwin began a second zoological notebook that was distinctly different from his scarlet notebook. Now he was a man with a clear intellectual project. The marine invertebrate notes in this journal show the marks of Grant's training and way of seeing. All Darwin's notes here concern the stages of zoophyte reproduction. Only six months before, Grant had published his discovery that the eggs of the sea sponge were free-swimming, propelling themselves along by the vibration of small hairs or cilia that covered their bodies, ending his essay with the words: 'How far this law is general with zoophytes must be determined by future observations.'[18] Now he had turned to zoophytes – organisms that were like the sea sponge – to see if he could determine whether these free-swimming eggs, bursting from the bodies of their parents like seeds, were indeed common to all zoophytes. In order to produce the evidence more quickly, he enlisted Darwin and Coldstream to work on the *Flustra*, a pale-brown seaweed-like zoophyte, known colloquially as 'sea-mat', growing in clusters of rocks close to the shoreline, and made up of hundreds of interdependent, connected polyps. Did this simple, primitive organism also ejaculate free-swimming ova like the sea sponge?

Grant, Coldstream and Darwin were all working closely together, by this point, dissecting in Grant's rooms and even on the beach itself, searching for the presence of swimming eggs in as many different apparently immobile sea creatures as they could. Outdoor dissection had become important for Grant when working on marine organisms that quickly deteriorated when taken out of their natural habitat. He taught Darwin how to dissect out of doors, using a microscope, and talked excitedly with him about his theories of the aquatic origins of life in a way that would have been more unguarded than he was prepared to be in his published work. Darwin claimed in his *Autobiography* that he was astonished by Grant's Lamarckian conclusions, despite the fact that he had already encountered these ideas in his grandfather's *Zoonomia*. But if he was astonished, he was also wildly curious about Grant's developing speculations.

Darwin's March notebook is full of detailed illustrations and copious notes describing his zoophyte dissections and search for free-swimming ova in zoophytes or other marine invertebrates he could find in the dredging boats at Newhaven and on the black rocks of Newhaven and Leith. He made a significant discovery on 20 March 1827. Knowing it to be important evidence to support Grant's theories, he rushed to the Prestonpans house with the news as soon as the dredger moored in the harbour:

Having procured some specimens of the Flustra carbocea (Lam:) from the dredge boats at Newhaven; I soon perceived without the aid of a microscope small yellow bodies studded in different directions on it. – They were of an oval shape & of the colour of the yolk of an egg, each occupying one cell. Whilst in their cells I could perceive no motion; but when left at rest in a watch glass, or shaken they glided to & fro with so rapid a motion, as at some distance to be distinctly visible to the naked eye [...] That such ova had organs of motion does not appear to have been hitherto observed either by Lamarck Cuvier Lamouroux or any other author: – This fact although at first it may appear of little importance yet by adducing one more to the already numerous examples will tend to generalize the law that the ova of all Zoophytes enjoy spontaneous motion.[19]

About the same time, Darwin made a second discovery, also described in his notebook:

One frequently finds sticking to oyster and other old shells, small black globular bodies which the fishermen call great Pepper-corns. These have hitherto been always mistaken for the young Fucus Lorius . . . to which it bears a great resemblance . . . . But on examining some others I found that this fluid, acquiring by degrees a vermicular shape, when matured was the young Pontobdella Muricata (Lam.) which were in every respect perfect & in motion.[20]

Darwin found swimming eggs in almost all the zoophytes he dissected, but the more evidence he brought to Grant, the more tensions seemed to arise between them – tensions that to Darwin seemed inexplicable. Grant gave a paper on the free-swimming ova of the *Flustra* to the prestigious Wernerian Society on 24 March 1827, illustrated by preparations and drawings. It was clearly indebted to Darwin's findings as well as his own. He also announced the discovery of cilia on the young of other zoophytes and that he had discovered the mode of reproduction of the sea leech *Pontobdella muricata*. Darwin gave a paper on the ova of the *Flustra* to the Plinian Society three days later, his very first scientific paper. He was eighteen.

Given that Grant had been rearing colonies of *Flustra* in his Prestonpans house and watching for evidence of free-swimming ova night and day for some months, it is unlikely that Darwin would have witnessed free-swimming ova *before* Grant. However, it does seem likely that Darwin's excited tales of witnessing swimming *Flustra* eggs in a watch glass on a dredging boat would have frustrated Grant, who was always careful to stage-manage his scientific announcements. Could it be that Darwin's 'discovery' propelled Grant to give his paper earlier than he had wanted? However, in July, when Grant published further information about the ova of the *Pontobdella*, he did acknowledge 'my zealous young friend Mr Charles Darwin of Shrewsbury, who kindly presented me with specimens of the ova exhibiting the animal in different stages of maturity'.[21]

The relations between Darwin and Grant had cooled. A note, written by one of Darwin's daughters and allegedly found in a bundle of papers in 1947, provides some explanation for this:

I then made him repeat what he had told me before, namely his first introduction to the jealousy of scientific men. When he was at Edinburgh he found out that the spermatozoa (?)/ ova (?) of (things that grow on sea weed)/ Flustra move. He rushed instantly to Prof. Grant who was working on the subject to tell him, thinking he wd be delighted by so curious a fact. But was confounded on being told that it was very unfair of him to work at Prof. G's subject and in fact that he shd take it ill if my Father published it. This made a very deep impression on my Father and he has always expressed the strongest contempt for all such little feelings – unworthy of searchers after truth.[22]

But these were also, perhaps thankfully, Darwin's last weeks in Edinburgh. Grant too would leave Edinburgh that summer: powerful advocates were recommending him for a chair at London University. Darwin was destined for Cambridge. His father had decided that, given that he was not going to qualify as a doctor, he should try for the Church. Perhaps in Cambridge he would show more application. Coldstream had decided to go to Paris to continue his medical studies and to follow Grant's example. It is likely that several graduating members of the Plinian Society set off for Paris in the summer of 1827. Perhaps they travelled together. Grant joined them there later in the summer for his annual visit to the Paris museums and libraries. William Browne, now beginning to specialize in the medical treatment of the insane, wanted to study in the Parisian asylums. Ainsworth, who would later become a surgeon and geologist to the Euphrates Expedition and achieve fame in documenting it, planned to study geology and medicine in Paris; and Darwin, though destined for Cambridge studies, also travelled to Paris in June and July of that summer. His Uncle Josiah was travelling to

Geneva to pick up his daughters, Fanny and Emma Wedgwood, who had been staying with an aunt for eight months, and he offered to take Charles as far as Paris. While his uncle travelled on to Geneva, Charles was left in Paris to occupy himself for several weeks. Given his invertebrate interests, he is likely to have spent some time with Coldstream and his other Plinian Society friends, accompanying them to the natural history museums and specialist libraries.

Darwin found Coldstream ill at ease, even that first summer of his Paris studies. Paris science was beginning to unsettle Coldstream's mind and his religious beliefs even more than Dr Grant had. William Mackenzie of Mission House in Passy, in whom Coldstream confided, recorded that 'he was troubled with doubts arising from certain Materialist views, which are alas! too common among medical students. He spoke to me of his doubts, and manifested anxiety on the subject of religion.'[23] If the ideas of Paris made Coldstream anxious, those he encountered in Germany brought him breakdown. Soon he was in a Swiss sanatorium. Convalescing, he made decisions. There would be no further continental study, no further materialist temptations. He would return to Leith in 1828 and work simply as a local doctor, healing the sick. He would shun the company of heretical zoologists such as Grant, Grey and Browne. 'In our day the majority of naturalists, I fear are infidels,' he wrote in 1829.

Coldstream, frightened by the sexual liberalism and hedonism of Paris and the temptations of materialist science, returned to Edinburgh in the middle of one of the most notorious trials of Scottish history: the trial of William Burke and William Hare for murder. Burke and Hare, two unemployed Irish navvies who had moved to Scotland to work on the Union Canal, had decided to reclaim unpaid rent from a deceased lodger by selling his body to Robert Knox, Dr Barclay's successor at Surgeons' Square, for £7/10s. A few weeks later they lured an old woman into their lodgings and suffocated her whilst drunk. This body, carried to Surgeons' Square in a wooden chest, fetched them £10. At the trial that began on Christmas Eve 1828, Burke pleaded guilty to supplying sixteen bodies to Robert Knox, but there may have been as many as thirty. Hare and his wife provided enough evidence to sentence Burke, but they were acquitted. Burke was hanged at the Head of Libberton Wynd at dawn on 28 January 1829, and his body taken to Dr Munro's dissecting theatre, where it was partially dissected before rioting crowds broke the windows of the dissecting rooms, demanding to see the body. Police were forced to allow groups of forty to fifty people at a time into the theatre. Men, women and children filed past the partially dissected body of the murderer as if it were the body of a king in state.[24]

Living back with his parents in Leith, Coldstream continued to be haunted by festering bodies and moral corruption. 'A fair exterior covers a perfect sink of iniquity,' he wrote of himself on 1 January 1830. He blamed his mental weakness on lack of discipline and temptations of the flesh: 'my present condition of mind is much inferior in strength and solidity to what it might have been had I not given loose reins to my lustful appetites. I have been ruined and enervated by a life of effeminacy and slothful indulgence.'[25]

Even this breakdown could not keep Coldstream from his marine creatures, especially now that he was living and working back on the shore of the Firth of Forth; but with the infidel Grant in London and the Revd John Fleming offering him non-heretical interpretations of marine zoology, Coldstream returned to some kind of peace even in his rock-pool hunting. He married in 1835 and settled down to family life and his growing medical practice. He continued to write occasional non-controversial marine pieces for Todd's *Cyclopaedia of Anatomy and Physiology* – on jellyfish (1835), limnoria (1834) and barnacles (1836).

From Cambridge, Darwin wrote to Coldstream in Edinburgh in 1829, expressing his sympathy for Coldstream's illness and grief that his friend had apparently given up marine zoology, arguing that 'no pursuit is more becoming for a physician than Nat. Hist.'[26] Coldstream agreed, but added in his letter that he had decided to dedicate himself to 'useful knowledge'. Natural philosophical speculations were not for the faint-hearted. He had also resolved to give up dissecting live creatures, for it seemed to be against nature.

Nevertheless he asked Darwin to pass on his regards to Dr Grant and ended: 'Be so good as to write me again soon, and tell me something of the present state of Natural History in Cambridge. Have you had any opportunity of studying marine Zoology since you left this?'[27]

Darwin had not had any experience of marine zoology since leaving Edinburgh. How could he? In Cambridge, eighty miles from the sea, his attention had turned to beetles. Professor Henslow, his new, less-radical botanist mentor and walking companion, was already remarking: 'What a fellow that D. is for asking questions.' Passing his exams in 1831, Darwin was now ready to take on a parish and settle down. Instead, his imagination full of the travelling tales of Alexander von Humbolt in South America, his thoughts turned to tropical expeditions. He wrote: 'It strikes me, that all our knowledge about the structure of our Earth is very much like what an old hen wd know of the hundred-acre field in a corner of which she is scratching.'[28]

The answer was to leave the corner of the field and scratch elsewhere, but what would his father have to say to this? In September 1831 the opportunity presented itself in the form of an invitation to accept the position of ship's naturalist on a map-making voyage of coastal South America on board the *Beagle*. His father, exasperated by his son's apparent flightiness, refused to agree to the voyage, but his Uncle Jos intervened, arguing that 'Natural History . . . is very suitable to a Clergyman', and eventually Darwin's father agreed.[29] Darwin wrote one last time to Coldstream, asking for advice and information about deep-sea dredging and techniques for meteorological observations. In particular he asked Coldstream to draw him a diagram of an oyster dredger in such a way that he could have one designed for the *Beagle* voyage. Coldstream drew careful instructions and diagrams and urged Darwin to contact Robert Grant in London for further advice.

Darwin bought one of the most powerful microscopes available in the world before he sailed, a microscope purchased from Bancks and Son of 119 New Bond Street. A botanist and physiologist, Robert Brown, who was also interested in zoophyte research, had recommended it to him. It was portable and especially designed with a watch glass fitted to the object stand to allow for submerged dissection.

Darwin and Grant met in London in the weeks before the *Beagle* sailed. There are no records of the meeting, only a series of notes Darwin made, recording preservation instructions given to him by his old teacher. What must Grant have felt, called upon by his old protégé? Grant, the great traveller, who had crossed European mountain ranges and walked Mediterranean coastlines in the footsteps of Aristotle, was now, without the luxury of a private income like Darwin, a long way from the sea, his days filled entirely with writing lectures and giving tutorials, struggling to make ends meet. Darwin came to him, twenty-two years old, asking for precise instructions on collecting marine creatures in exotic, tropical seas.

3

A Baron Münchausen Amongst Naturalists

*

We may begin where we please,
We shall never come to an end;
our curiosity will never slacken

George Henry Lewes, *Studies in Animal Life,* 1862

It is 6 January 1832 – nightfall. The *Beagle* has been at sea for two weeks. She has sailed past the Dry Salvages and Madeira, and now she is moored off the coast of the Canary Islands. Darwin has been terribly seasick and today has been one of the first times he has ventured on deck; now he sits under the stars, writing his diary, gazing across to the lights of the port and trying to make out the shapes of the island's volcanic cones in the darkness. Confined to his hammock below deck, he has been dreaming of the volcanic wonders of 'the peak of Tenerife and the great Dragon tree; sandy, dazzling, plains, and gloomy silent forest'[1] of the Canary Islands. It is, for Darwin, 'perhaps one of the most interesting places in the world'.[2]

Now he can see the lights of Tenerife across the bay – within reach; but the crew are not allowed to land. There is a cholera epidemic sweeping across Britain and the consul in Santa Cruz has imposed strict quarantine regulations on all British ships – twelve days' quarantine. Captain FitzRoy, though mortified on behalf of his young guest, cannot wait that long and has given orders to sail south. The *Beagle* has work to do and a schedule to maintain. The coastline of South America has to be mapped, charts are to be checked and extended, tides and weather conditions measured and recorded constantly on this voyage around the world. Darwin is devastated. He feels the prohibition like a 'death warrant'.[3] Yet, the beauty of the sea and the tropical night are consolations of the most exquisite kind: 'The night does its best to soothe our sorrow – the air is still & deliciously warm – the only sounds are the waves rippling on the stern & the sails idly flapping round the masts – Already I can understand Humboldt's enthusiasm about the tropical nights, the sky is so clear & lofty, & stars innumerable shine so bright that like little moons they cast glitter on the waves.'[4]

There is nothing to do; the ship is becalmed. The other sailors and officers who are not on duty sit on deck. Some play cards, others write their first letters home. A group of three young, well-dressed and portly Fuegians sit together watching the sea, talking in broken English. Captain FitzRoy, who brought them to England the year before, has promised to return them to their families in Tierra del Fuego in the care of a young missionary called Richard Matthews. The *Beagle* has thus carried four Fuegians to England to be educated and converted to Christianity, Fuegians given pantomime names by the officers: Fuegia Basket, York Minster, Jemmy Button and Boat Memory. In Britain they have been educated in a private boarding school, examined by phrenologists, instructed in the 'simple arts of civilised life'[5] and presented to King William and Queen Adelaide at court. Only three are returning: Boat Memory died in the naval hospital of Plymouth after a smallpox vaccination, a dose too heavy for his small body.

The sailors and officers are not entirely at ease with Darwin yet. Pecking orders and territories are still being established and Darwin's chronic seasickness does not bode well. Employed as gentleman companion to the Captain, he is the only man to share the Captain's table, yet, given FitzRoy's unpredictable moods, this is a rather dubious honour. In such a small community the emotions and moods of an individual, especially the captain, will affect everyone. A previous captain of the *Beagle*, fearing breakdown, had shot himself in the captain's cabin only four years before. There are jealousies, too, even this early in the voyage. The ship's surgeon, Robert McCormick, an experienced naturalist, is suspicious and jealous of Darwin's dissecting station in the poop cabin. Traditionally, it is the surgeon's job to be the ship's naturalist on an Admiralty voyage. It has always been his responsibility on his previous voyages. The ship's crew has a rigid hierarchy; yet, financially independent,

Darwin exists outside it. He will have to use his good nature and charm to establish his own territory.

Darwin goes below deck to make the most of the last hour or two of light before lights out, grasping the handrail as the ship rolls to one side and steeling himself against a rising wave of nausea. Captain's orders are clear: ship's candles are to be extinguished by nine o'clock in harbour, half past eight at sea.[6] During the day Darwin shares this poop cabin with the ship's nineteen-year-old mate and assistant surveyor, John Lort Stokes, and the fourteen-year-old midshipman Philip Gidley King, who use the large chart table in the middle of the room to work on the surveying documents. There is no time to be lost and he has a new golden rule to live by: *Take care of the minutes*.[7] In the evenings Captain FitzRoy sometimes invites the ashen-faced and nauseous Darwin into his cabin for conversation and, to their mutual pleasure, they have discovered they share an admiration for the novels of Jane Austen.

A crew of seventy-five young men sailing around the world, most in their twenties, like Darwin and the Captain; one woman: the teenage Fuegia Basket, already 'betrothed' to York Minster, who is jealous of her every movement. Several of the young sailors have an interest in natural history: McCormick, the ship's surgeon, and Benjamin Bynoe, the assistant surgeon, are well read in botany, zoology and geology and are putting together their own natural-history collections; 1st Lieutenant Wickham is a plant collector and 2nd Lieutenant Sulivan has geological interests. The ship's steward, Harry Fuller, and the ship's clerk, Edward Hellyer, collect birds. Even the boy, Philip Gidley King, is interested in marine invertebrates, for his father, Philip Parker King, the Captain of the first *Beagle* survey, keeps a rare barnacle collection on his estate in Australia.

Darwin is already aware that the lack of distractions, the strict regimes imposed by FitzRoy and the cramped conditions he has for his work are nurturing in him a discipline that is new and welcomed, a natural progression from the training Grant imposed. When he writes from the boat to his older sisters, who since his mother's death have nurtured his moral and educational development and mediated between him and his father, he tells them how well he is working: 'I find to my great surprise that a ship is singularly comfortable for all sorts of work. – Everything is so close at hand, & being cramped makes one so methodical, that in the end I have been a gainer . . .'[8] The poop cabin is very small indeed – ten feet by eleven feet – so that at night Darwin and the boy midshipman King must hang up their hammocks across the tables, piles of books, measuring equipment and maps. They watch the movements of the stars through the skylight above them. The heavy furniture creaks slightly with the roll of the waves and at night they can hear the footsteps of the night sentinel on the deck above. Stokes sleeps in a bunk bed under the stairs just outside the door. There are hundreds of volumes here: atlases, dictionaries, Bibles, novels, travel narratives, books on zoology, volcanoes, mineralogy and Darwin's precious copy of John Milton's *Paradise Lost*. Captain's orders: all officers are permitted to borrow books from the poop cabin; books must be signed out and returned by 8.30 p.m., lights out.

Darwin's candle casts a pool of light upon his part of the table. Polished mahogany gleams in the candlelight and shadows play across the walls as the ship rolls in the swell. Collecting drawers, especially made for him by the ship's carpenter, make the most of the restricted space, but two weeks into the voyage the drawers are still empty. A dead black-backed gull lies on the table next to Darwin's portable microscope, books, dissecting tools and a glass jar full of slightly milky seawater. It gives out a faint green glow in the darkness of the cabin. He examines a drop of the water under his microscope, pauses for a moment and takes his pen to begin the very first of his *Beagle* zoology notebooks:

January 6th 1832: The sea was luminous in specks & in the wake of the vessel, of a uniform, slightly milky colour. – When the water was put into a bottle, it gave out sparks for some minutes after having been drawn up. – When examined both at night and next morning, it was found full of numerous small (but many bits visible to the naked eye) irregular pieces of (a gelatinous?) matter. The sea next morning was in the same place equally impure.[9]

Zoophytes – luminous zoophytes. Not quite plants, not quite animals, these swarming organisms, invisible to the naked eye, lit up the sea and sky. They were also, he discovered, in the strange red dust that fell upon the boat from time to time. Christian Gottfried Ehrenberg, the leading European microscopist, who was working on infusoria in Berlin and who had travelled with Alexander von Humbolt in Siberia, was using infusoria dissections to question Cuvier's fixed and hierarchical system of classification. To his mind there were no such things as superior and inferior organisms – all had the organs necessary for the full range of animal activities: 'The infusorian has the same sum of organisation-systems as a man,' he had written in 1835.[10]

Next Darwin turns his attention to the black-backed gull. Frustrated by Tenerife's quarantine sanctions, he had taken his pistols up on to deck that afternoon. It is an unremarkable bird in itself, not worth collecting, but Darwin is only interested in the contents of its stomach, for he is looking for new marine invertebrates. The plumage is soft in his hand as he inserts the dissecting knife. The contents of the bird's stomach spill out on to the paper laid carefully across the wood. He puts the bird to one side to examine the pile of warm, half-digested molluscs and scraps of fish, most now quite unrecognizable. Poking at the pile with his dissecting knife, he finds a small cuttlefish here, still intact, about two inches long. Its bulbous eyes perhaps still register the surprise of being rudely snatched from tropical waters and the wonder of its short moments of flight before being swallowed whole by its feathery predator in mid-air. It is not a species that Darwin recognizes. Perhaps Professor Henslow will be able to identify it. This will be Specimen 1 of Darwin's *Beagle* collection. Dried, wrapped in paper, labelled and tucked inside a box within a crate, it will be posted to Cambridge a few months later. Darwin's *Beagle* collection, five years in the making, begins with a cuttlefish. His first zoology notes describe phosphorescent zoophytes.

Only weeks later, walking the hot, rocky beach of St Jago, a volcanic outcrop in the middle of the Atlantic, 450 miles from the coast of Africa, he found a small octopus marooned in a rock pool; but before he was able to reach it with his net, it darted into a crevice in the rock, leaving the water in the pool discoloured by a dark chestnut-brown ink. Darwin had learned patience hunting in the rock pools of Scotland. He could wait. He could be very still. Grandmother's Footsteps. He watched the octopus emerge from its crevice as the water cleared again. But now he could hardly see it. Chameleon-like, it had changed its colour, disguising itself before his very eyes to match the colour of the rock-pool floor: French Grey with minute spots of yellow, disappearing and appearing by turns. Fascinated, rooted to the spot, he could not even reach for his net. Clouds of colour, Hyacinth Red and Chestnut Brown, continuously passed across its body. But were they like clouds or blushes? Darwin wondered, struggling with analogies. Putting these creatures into words taxed his vocabulary and his imagination.

But who was hunting whom? As Darwin watched, the octopus remained 'for a time motionless, it would then advance an inch or two, like a cat after a mouse, sometimes changing its colour'. Then it would dart away, 'leaving a dusky train of ink to hide the hole into which it crawled'.[11] Frustrated and amused, Darwin turned away, turning his back on his prey, wading further into the pools to look for other marine animals; but he was constantly distracted by the sound of grating behind him. The octopus he had abandoned was squirting water in his direction, aiming directly at him from its rock pool. Following the line of spray stealthily, keeping his head down, Darwin surprised his octopus playmate with his net and slipped it into his collecting jar. This animal had to be observed more closely. Back in the cabin that night, the ship at sea and Darwin sick again, the chameleon octopus glowed angrily in the dark, the only light in the pitch-black cabin. It did not live long. Darwin was too sick to either release it or label it. 'Here I have spent three days in painful indolence, whilst animals are staring me in the face, without labels or epitaphs,' he complained on 12 February 1832.[12]

Grant's zoophytes and swimming eggs still haunted him here right from the start of the voyage. Becalmed off Tenerife, he had seen the sea lit up by zoophytes much too small to see. This was an opportunity not to be missed, an opportunity to study the body structures of these luminous borderline creatures. The plankton net he had made modelled on Coldstream's diagram of an oyster dredger from the Firth of Forth worked well: 'I proved today the utility of a contrivance which will afford me many hours of amusement & work – it is a bag four feet deep, made of bunting, & attached to [a] semicircular bow this by lines is kept upright, & dragged behind the vessel – this evening it brought up a mass of small animals, & tomorrow I look forward to a greater harvest.'[13]

Many of the books he had consulted with Robert Grant and John Coldstream were in the ship's cabin – expensive, leather-bound books on marine invertebrates by Lamouroux, Cuvier, Lamarck, Thompson and Fleming, most purchased from Treuttel, Wurtz and Richter, a specialist bookshop at 30 Soho Square, London.[14] Sometimes he would write guiltily home to his sisters to ask them to buy new books from Treuttel's and post them out to him: 'When you read this I am afraid that you will think that I am like the Midshipman in Persuasion who never wrote home, excepting when he wanted to beg: it is chiefly for more books; those most valuable of all valuable things . . .'[15] But there were many species he found that were still unmapped, even in the newest zoology books they sent.

The first day he used his plankton net, it brought in a jellyfish and Portuguese man-of-war, *Physalia physalis*. The man-of-war looks like a jellyfish, but its jelly-like pale-purple translucent float has a comb of pleats along the top like a Mohican haircut. This sea punk uses its raised crest like a sail to catch the wind. Beneath the surface of the water, its trailing tentacles, some tens of metres long, look like chains of coiled purple jewels. But this is no *man*-of-war; it is a veritable army, a colony of highly poisonous polyps, dependent upon one another for survival. Under the float one community of polyps breeds and dies, while the deadly tentacles of others catch and paralyse small fish. Another group of polyps seizes this food and spreads over it to digest it – one body, one colony of bodies, hunting, breeding, being born, digesting, dying. Where does life begin and end here? Where does an individual begin and end?

Darwin, excited, tried to tease out the pale purple tentacles caught up in the dredging net: 'getting some slime on my finger from the filaments it gave considerable pain, & by accident putting my finger into my mouth I experienced the sensation that biting the root of the Arum produces.'[16] Perhaps only the obsessively curious Darwin could have made such a comparison. Had he experimented with the arum root too in the gardens at Shrewbury as a small boy? Knowing it to be poisonous, had he dug up the potato-like root and bitten into it, just to be sure? Just to feel the painful sting on his tongue. Just to know.

WHITES.

Nº	Names.	Colours.	ANIMAL.	VEGETABLE.	MINERAL.
1	Snow White.		Breast of the black headed Gull.	Snow-Drop.	Carara Marble and Calc Sinter.
2	Reddish White.		Egg of Grey Linnet.	Back of the Christmas Rose.	Porcelain Earth.
3	Purplish White.		Junction of the Neck and Back of the Kittiwake Gull.	White Geranium or Storks Bill.	Arragonite.
4	Yellowish White.		Egret.	Hawthorn Blossom.	Chalk and Tripoli.
5	Orange coloured White.		Breast of White or Screech Owl.	Large Wild Convolvulus.	French Porcelain Clay.
6	Greenish White.		Vent Coverts of Golden crested Wren.	Polyanthus Narcissus.	Calc Sinter.
7	Skimmed milk White.		White of the Human Eyeballs.	Back of the Petals of Blue Hepatica.	Common Opal.
8	Greyish White.		Inside Quill-feathers of the Kittiwake.	White Hamburgh Grapes.	Granular Limestone.

6 Darwin's Colour Chart

The colours of the sea creatures, like the gorgeous purples and pinks of the man-of-war, amazed him. They seemed to get more vivid and gaudy as the ship sailed further south.[17] By February and March he was struggling for words to describe the colours of the marine creatures he was cataloguing daily. One of his zoological friends had insisted that he take the geologist Abraham Gottlob Werner's colour charts with him, usefully edited into a book by the Edinburgh miniaturist Patrick Syme.[18] The colour charts enabled scientists to match the colour of a cuttlefish tentacle or piece of granite or butterfly wing to a named colour that could be consulted by other naturalists. To indicate that he was using Werner's colour chart names, Darwin capitalized them in his descriptions: Lake Red. Ink Black. China Blue. Saffron Yellow. Peach Blossom and Aurora Red. Pistachio Green. Oil Green. Wood Brown – colours to taste, smell and touch; colours in poetry, colours spanning the vegetable mineral, and animal kingdoms:

Skimmed Milk White	White of human eyeballs	Back of the petals of Blue Hepatica	Common Opal
Wine Yellow	Body of the Silk Moth	White Currants	Saxon Topaz
Flesh Red	Human Skin	Larks Spur	Heavy Spar Limestone

It was overwhelming. These tropical seas yielded sensual pleasures beyond his imaginings – beyond even the treasures of Prestonpans. Sitting amongst the rock pools on the western rocky shore of Quail Island in the Atlantic, eating ripe tamarinds and biscuits towards the end of January, Darwin remembered the sands of Leith:

Often whilst at Edinburgh, have I gazed at the pools of water left by the tide; & from the minute corals of our own shore pictured to myself those of larger growth: little did I think how exquisite their beauty is & still less did I expect my hopes of seeing them would ever be realised. – And in what a manner has it come to pass, never in the wildest castles in the air did I imagine so good a plan; it was beyond the bounds of the little reason that such day-dreams require. – After having selected a series of geolog. specimens & collected numerous animals from the sea, – I sat myself down to a luncheon of ripe tamarinds & biscuit; the day was hot, but not much more than the summers of England.[19]

Would people believe his accounts of these bizarre and gaudy creatures? he wondered. They would sound so fantastic. He wrote to Henslow later that year: 'After this I had better be silent. – for you will think me a Baron Münchausen amongst Naturalists.'[20]

In 1785 a jewel thief from Hanover named Rudolf Erich Raspe published a satirical book in England, *The Adventures of Baron Münchausen*, which claimed to be based on the life and travels of a notorious Baron Münchausen, who had a reputation for embellishing his war stories. The book included, for instance, the Baron's 'Extraordinary flight on the back of an eagle, over France to Gibraltar, South and North America, the Polar Regions, and back to England, within six-and-thirty hours' and told of the time the baron tethered his horse to a 'small twig' in a snowstorm, and discovered when the snow melted that the twig was actually a church steeple. Were Darwin's theories castles in the air like the Baron's? His imagination didn't torment him as Coldstream's had – his only problem was how to prove such castles in the air, how to bring them into the realm of plausibility. That's what all the scratching and recording and collecting of evidence was for.

7 Botofogo Bay

The heat was so extreme as the ship sailed towards Rio de Janeiro in February 1832 that Darwin slept naked on the mahogany dissecting table at night. Even then he complained that he felt as if he were being stewed in warmed melted butter.[21]

Back in Leith the winter weather was not so balmy. John Coldstream, the doctor who still walked Leith sands in the early morning with his glass jars and nets, was also preparing for the cholera epidemic that was moving north, driven by the as-yet-unknown water-borne cholera bacillus. The disease had already killed 6,000 people in London. Most people who died did so in less than twenty-four hours, from dehydration caused by chronic diarrhoea. Coldstream and the other Leith medical men could do nothing but wait, order the medications from London and prepare a hospital for the sick. On Sunday, 22 January 1832, Coldstream wrote in his journal: 'The cholera is now within four miles of us and we are hourly expecting it at Leith.'[22] But, although cases were multiplying fast in the seaside village of Musselburgh, four miles away, the disease didn't arrive in Leith until the end of February. Coldstream knew that this was the beginning of chaos. On 29 February he wrote:

How various is the lot of men! Tonight all is quiet around me. I have no patients very seriously ill; all our family are well and happy; the town is peaceful and still, while only four miles off, pestilence prevails; men die by twenties and thirties a-day – all classes are panic-struck, and the medical men are labouring in their vocation by night and by day. How soon the aspect of things may be changed here, God only knows.[23]

Darwin's sister Susan wrote to him anxiously about cholera in Shrewsbury; there were already twenty cases in Darwin's home town by August 1832.[24] But it was the rapidly expanding industrial cities where cholera had taken its most secure hold, cities in which overcrowding and unsanitary conditions provided an ideal environment for the survival and propagation of the cholera virus but not for the survival of the urban poor. Britain's population had doubled in thirty years. The cholera pamphlets, many written by doctors for the middle classes, warned about the dangers and pestilence of the slums. Some even argued that the centres of the disease, so often the poorest areas of the town, such as the port areas of Leith, should be sealed off and policed. The middle classes flocked to seaside resorts in the hope that the sea air would dispel the supposedly airborne germs, not understanding that it was the water they drank that poisoned them. By September of that year, Coldstream, working night and day in Leith hospital since January, was suffering from chronic exhaustion. When his doctor ordered a seaside convalescence, he took lodgings in Torquay with two of his sisters and a cousin and devoted himself to a winter of marine zoology.

Cholera was sweeping across Ireland, too, as Darwin sailed ever closer to the shores of South America in 1832. In Cork an ex-army surgeon, William Vaughan Thompson, who had been promoted to the position of Inspector General of Cork Hospital, had just published a book called *Cholera Unmasked*. However, back in Britain it wasn't this doctor's work on cholera that people were talking about, but his work on barnacles, crustacea, and phosphorescent sea creatures, a book published in 1830 as *Zoological Researches and Illustrations; Or a Natural History of Nondescript or Imperfectly Known Animals*. This series of memoirs, based on painstaking microscopal observations, made him the talk of the zoological societies in London. Linnaeus and Cuvier had defined barnacles as molluscs through a lack of understanding of their metamorphosis and of their internal anatomy. Thompson had watched these metamorphoses through long hours at his powerful new microscope and proved that, like insects, they metamorphosed in extraordinary ways. He had seen free-swimming barnacles shape-shift into adult barnacles that glued themselves to rocks.

The philosophical implications of Thompson's discoveries were unlimited. If barnacles and crabs had been seriously misunderstood by Linnaeus, Cuvier and even Lamarck, he implied, every marine invertebrate had to be rethought. Dramatic body changes like these were supposed to be common only amongst insects. What kind of mirroring was this? Land and sea: patterns above and below the water. Thompson was a zoology detective working like Grant with patience, a powerful microscope and a watch glass. He might have called his zoology book *Barnacles Unmasked*, but, unlike Grant, he wasn't interested in the philosophical implications of his discoveries. He was satisfied to have discovered these underwater metamorphoses and to argue for reclassification. Barnacles were simply in the wrong boxes and needed to be moved to the right boxes, restored to their proper God-given family. Others, however, would use his amassed facts and observations for philosophical ends.

Thompson, like Darwin who ordered a copy of this groundbreaking book for the *Beagle* library before it sailed, was enthralled by the poetry of marine life forms. He was no dry systematist. He had sailed the world as a ship's doctor and, apart from the extraordinary descriptions of the metamorphosis of barnacles, his book was full of engaging accounts of phosphorescent tropical seas such as those Darwin had seen off the coast of Tenerife:

Returning from a fishing party late in a still evening across the bay of Gibraltar, in a direction from the Pomones river to the old Mole, in company with Dr Drummond (now Professor of Anatomy to the Belfast institution) and a party of naval officers, the several boats, though separated a considerable distance, could be distinctly traced though the gloom by the snowy whiteness of their course, while that in which we were, seemed to be passing through a sea of melted silver; such at least was the appearance of the water, displaced by the movement of the boat and the motion of the oars, the hand, a stick, or the end of a rope, immersed in the water, instantly became luminous and all their parts visible, and when withdrawn, brought up numerous luminous points less than the smallest pin's-head, and of the softest and most destructible tenderness.[25]

Darwin, too, continued to see swarms of phosphorescent zoophytes, no bigger than a pin's head, lighting up the surface of the sea like glow-worms, wherever they sailed. On route to Buenos Aires in October 1832:

The night was pitch dark with a fresh breeze. – The sea from its extreme luminousness presented a wonderful & most beautiful appearance; every part of the water, which by day is seen with foam, glowed with a pale light. The vessel drove before her bows two billows of liquid phosphorus, and in her wake she was followed by a milky train. – As far as the eye reached, the crest of every wave was bright; & from the reflected light, the sky just above the horizon was so utterly dark as the rest of the heavens. – it was impossible to behold this plain of matter, as it were melting and consuming by heat, without being reminded of Milton's description of the regions of Chaos and Anarchy.[26]

Under the microscope these phosphorescent 'lower' creatures were beautiful and strange, he wrote: 'Many of these creatures so low in the scale of nature are most exquisite in their forms & rich colours. – It creates a feeling of wonder that so much beauty should be apparently created for such little purpose.'[27]

So much beauty for so little purpose. Unlike John Vaughan Thompson, Darwin did not try to frame his discoveries in terms of a revelation of God's benevolent purpose. Thompson really believed that the zoophytes' purpose, the reason they had been put on the earth, was to light up the ocean at night so that sailors would be able to avoid shipwreck. 'The object of the Creator is not always obvious,' he concluded.[28] Though Darwin was not a deist like his father and grandfather, he had learned from them the habit of separating his religious belief from his scientific enquiry. It wasn't that he didn't feel a divine presence in nature – in the vast beauty of a tropical forest or under a starlit sky – but that framing all the questions that dogged him in terms of the purpose of a divine creator just was not a habit of mind. When he found Thompson searching for these explanations above all other possible explanations, it surprised him. What was interesting to him now was precisely the apparent *lack* of purpose of the zoophytes' nocturnal lights. Was it a signal to attract the opposite zoophyte sex? It couldn't be, because most zoophytes were hermaphrodite. The beauty, Darwin was sure, was simply a by-product of nature's processes: the phosphorescence, he believed, was caused by decomposing bodies of millions of dead zoophytes amongst the live ones, a process by which the ocean purified itself – like breathing out toxins. It was purpose enough; and poetry enough.

Elsewhere, other swarms of infusoria or confervae, filaments of microscopic plants, made the surface of the sea look red or muddy brown. Off the coast of Chile, Mr Sulivan brought Darwin a watch glass filled with seawater stained a pale red. He had already taken a look at the water under his own microscope and had seen, he said, 'moving points'. They took turns looking through the lens of Darwin's stronger microscope. Darwin saw thousands of oval shapes, contracted with a hairy ring around their middle, hairs that the organisms were using to propel themselves through the water. The instant they stopped moving their bodies burst. Simply exploded, ejecting coarse brown granular matter into the water.[29] South of Bahia, in September 1832, the whole surface of the sea was discoloured with bands of mud-coloured water, sometimes miles long. Under the microscope these confervae looked like chopped bits of hay. With a stronger lens Darwin could see that each part was made up of bundles of between twenty and sixty cylinders, and sacks filled with granular matter.

Where did they come from and what was their purpose? Had they been spontaneously generated, these animalcules and confervae? Some people, such as his grandfather Erasmus Darwin, argued that all life began from such aquatic fragments. Had they entered the world without parents? If not, where was their parent body? Were they plants or animals? What kept them together in such dense colonies so that the edge of the strips of discoloration were so clearly marked from the rest of the sea?

By this point in the voyage Darwin was preoccupied with geological puzzles as well as zoological ones. The two had become impossible to separate, now that he was gathering information about how the Earth had begun. Shellfish on the tops of mountains told of early forms and moving seas and bodies sealed and imprinted in rocks. Charles Lyell's new book *Principles of Geology* had gripped his imagination and the imaginations of geologists all over the world. Lyell was interested in volcanic landforms and raised beaches. He was puzzling out the slow, drip-on-drip formation of the Earth's surface over eons of time and arguing that it had been formed not by sudden, violent spasms of eruption and catastrophe, as others had argued, but through forces still working in the world: river, rain, rising seas, volcanoes and wind. For the Earth to have been formed this way, it had to be millions of years old, not the presumed 4,000 years of biblical history, and land masses were still moving, invisibly metamorphosing. Wherever he travelled, Darwin found himself looking at mountain ranges and seashores through Lyell's eyes, seeing through Lyell's theory.

Of the pages of Darwin's zoology notebooks, written during his five years on board the *Beagle*, over half concern marine inverte-brates. Over a third of the notebooks concern zoophytes.[30] For Darwin and other zoologists working across Europe they were an alternative way of thinking through the beginnings of time, raising philosophical questions shared by the study of geology. In May 1833, Darwin wrote to William Darwin Fox: 'The invertebrate marine animals are . . . my delight; amongst them I have examined some, almost disagreeably new; for I can find no analogy between them & any described families.'[31]

8 Repairing the *Beagle*

Whenever he was at sea he collected zoophytes and other creatures in his plankton net. When Captain FitzRoy ordered the *Beagle* to be beached in April 1834 near Santa Cruz, so that the ship's carpenter could repair the ship's keel and check for barnacle damage – a job he had to complete in a single tide cycle – it was Darwin who scoured the coppered underbelly of the ship for interesting marine invertebrate hitchhikers to slip into his collecting jars. Often the invertebrate dissections would continue for days at a time. Whilst a description of the dissection of a beetle or frog might take up a few lines, the dissection of a single zoophyte often ran to six or eight pages.[32] 'Most assuredly', he wrote to Henslow, 'I might collect a far greater number of specimens of Invertebrate animals if I took less time over each.'[33] Much of this intensive research remained unpublished.

With his powerful microscope Darwin could see further into his sea creatures' minute bodies than almost anybody in the world had ever done before – new bodies; new worlds inside them. Now Darwin could not only continue to search for swimming eggs in simple organisms; he could perhaps even move inside the swimming eggs. He found small, moving, vibrating darting structures in the tissues of all the marine animals and plants he dissected. Darwin called these minute particles 'granules' or 'granular matter' or 'grains', and he was sure they had something to do with reproduction. Robert Brown, the zoophyte expert, called them 'active molecules'. The trouble was not just with the names but with the explanations.

By the end of the *Beagle* voyage, Darwin was sure that this granular matter was primordial; these grains were the building blocks of all organic life, animal and vegetable, from algae to elephants – unity in extreme diversity. The grains were common to both kingdoms, animal and vegetable. The kingdoms could be united here at this microscopic point, this common denominator. In addition, his microscopic work had proved to his satisfaction that in both kingdoms both asexual and sexual reproduction took place.[34] They were kin. There were more and more points of comparison. By the end of the *Beagle* voyage he was convinced that: 'All animals of the same species are bound together just like buds of plants, which die at one time. Though produced either sooner or later.' The many buds of one parent tree; common ancestry. He made an important note to himself in his 1837 transmutation notebook on his return: 'Prove animals like plants; trace gradation between associated & non associated animals. - & the story will be complete.'[35] Yet it would be some time before Darwin began to question this assumed hierarchy in nature, a great chain of being. It was a habit of thought almost impossible to resist because it dominated the ideas of most of the philosophical naturalists, however radical. There were higher animals and lower animals and the job of the lower animals was to serve the higher animals as food or as fuel, or even as ocean lights: each to his place and purpose.

Darwin's marine zoology research on the *Beagle* demonstrated again and again how interdependent and symbiotic life was. Every living marine organism had a part to play in supporting the life of other species. Just as he marvelled at what we might call the 'eco-system' of the tropical forests of Tierra del Fuego, so he found giant aquatic forests growing on all the submerged rocks around the coast of the Falkland Islands and Tierra del Fuego. Sailors navigated around these kelp forests, which warned of dangerous rocks beneath; but the fact that these forests not only saved the lives of sailors, by acting as lighthouses, seemed inconsequential compared to the fact that they also supported hundreds of species of corals, polyps, and sea anemones: 'On shaking the great entangled roots, a pile of small fish, shells, cuttle-fish, crabs of all orders, sea-eggs, star-fish, beautiful Holuthuriae . . . Planariae, and crawling nereidous animals of a multitude of forms, all fall out together.'[36]

Yet this agent of life could also be an agent of death. In March 1833 the ship's clerk, Hellyer, was reported missing. He had gone ashore on to the Falkland Islands to shoot birds, for he was building up a collection of rare bird skins that he would have stuffed on his return to England. The search party found his clothes, gun, watch and a pile of dead birds on the rocky beach. But they were too late. They found his naked body as the tide retreated, half-submerged in the water. Swimming to retrieve a shot bird that had fallen into the water, his feet had become entangled in the kelp. His drowned and swollen body had to be cut out of the seaweed. These aquatic forests were both fertile and dangerous.[37] The grave of the man drowned by seaweed is still visible on a desolate outcrop of rock on the Falkland Islands.

Without the kelp, however, the marine food cycle of this particular coastline would be broken. Without shelter the fish and other marine creatures that lived in the giant kelp forests would not flourish. Without these creatures, cormorants, divers, seabirds, otters, seals and porpoises would die; and shoreline man was entirely dependent on this food chain too. It marked out his separation from cannibalism and from extinction. Without it, 'the Fuegian savage, the miserable lord of this miserable land, would redouble his cannibal feast, decrease in numbers, and perhaps cease to exist'.[38]

Darwin was both fascinated and disgusted by the Fuegians. Half-naked, they scratched a miserable existence on the rainy shores of Tierra del Fuego, living on fish and limpets. These were shore people who depended entirely on food from the sea, food nurtured in the kelp forests. From the deck of the *Beagle*, in February 1834, he watched a colony of Fuegians living on the beach: 'Here 5 or 6 human beings, naked & uncovered from the wind, rain & snow in this tempestuous climate sleep on the wet ground, coiled up like animals. – In the morning they rise to pick shell fish at low water; & the women winter & summer dive to collect sea eggs; such miserable food is eked out by tasteless berrys & Fungi.'[39] Life was in a state of complex entanglement, yet the practice of so many zoologists was to separate organisms from their environment and from the other organisms with which they were entangled and to box and pin them in splendid isolation.

It seemed to be something about the Fuegians' exposure to the elements that disgusted and disturbed Darwin. He wrote that he had never felt the division between a savage and a civilized state so acutely as in watching the Fuegians, particularly watching the women in the rain:

But these Fuegians in the canoe were quite naked, and even one full-grown woman was absolutely so. It was raining heavily, and the fresh water, together with the spray, trickled down her body. In another harbour not far distant, a woman, who was suckling a recently-born child, came one day alongside the vessel, and remained there whilst the sleet fell and thawed on her naked bosom, and on the skin of her naked child. These poor wretches were stunted in their growth, their hideous faces bedaubed with white paint, their skins filthy and greasy, their hair entangled, their voices discordant, their gestures violent and without dignity.[40]

This was no South Seas golden-sands idyll complete with flower garlands and conch shells. The beach the Fuegians lived on was covered with boulders, and because the cliffs were so steep in places, they could only move about in their canoes. They had no homes, he noted, and no domestic affection. They lived like animals. But they had survived – and continued to survive – here in this inhospitable, cold and wet place. Whatever his own judgements, Darwin was aware that their survival here meant that they had successfully adapted to their environment: 'Nature by making habit omnipotent, and its effect hereditary, has fitted the Fuegian to the climate and the productions of his country.'[41] No one knew what effect the transplantation of the 'civilised' Fuegians would make on these communities.[42]

Darwin and the other officers on the ship also had to adapt to their changing environment. Frock coats and clean-shaven faces would not help them survive against the biting insects and elements. They grew beards before they sailed for Patagonia.[43] That August, Darwin wrote to a friend from Montevideo: 'If you were to meet me at present I certainly should be looked at like a wild beast, a great grisly beard and flushing jacket would disfigure an angel.'[44] He ate roast armadillos on tropical shores, swam in warm seas, slept in makeshift tents on mattresses of putrefying seaweed on rocky coasts, and slaughtered scores of seabirds with his geological hammer. In Tahiti he chewed on hallucinogenic plants so powerful that even the local people wouldn't touch them.[45] 'With my pistols in my belt and geological hammer in hand, shall I not look like a grand barbarian?' he asked in 1832.[46]

There were cultural and religious rituals, too, that had to be accommodated in the most bleak conditions. On Christmas Day 1834 twelve men of the ship's company, including Darwin and Sulivan, foraged for birds' eggs on a small island in the Chonos Archipelago off the coast of Chile in the rain. They needed enough eggs to make a plum pudding. Breaking into the padre's house attached to the church, they hung up their clothes to dry and, half-naked, roasted a sheep and made two enormous plum puddings. Lieutenant Sulivan told the story in a letter home:

It would have amused you if you could have seen us in a dirty room with a tremendous fire in the middle, and all our blankets and clothes hung round the top on lines, getting smoked as well as dry, while all hands were busily employed for four hours killing a sheep, picking raisins, beating eggs, mixing puddings which were so large that, in spite of two-thirds of the party being west-country men, we had enough for supper also.[47]

That Christmas in 1834, Darwin had found the Chonos Archipelago one of the most beautiful places in the world: 'I cannot imagine a more beautiful scene, than the snowy cones of the Cordilleras seen over an island sea of glass, only here & there rippled by a Porpoise or logger-head Duck.'[48] Cones mirroring cones.

Lowe's Harbour, Chonos Archipelago, January 1835. Darwin, twenty-six years old, walked along a sandy beach covered with small shells on an island off the coast of southern Chile. The *Beagle* was moored in the harbour nearby. Lush forests lined the beach where he walked, hanging down like an evergreen shrubbery over a gravel walk. In the distance he could see the four great snowy cones of the range of mountains called the Cordillera. There were clumps of wild potatoes growing nearby on the sandy, shelly soil near the beach, and out in the bay sea otters hunted small red crabs, which swam in shoals near the surface of the water. Here on the Chilean beach he turned over another object worth adding to his collection: a thick conch shell shot through with hundreds of unusual and tiny boreholes. This shell would become one of his most prized possessions. However, it was the holes in the shell rather than the shell itself that interested him. What creature had made them and why? He peered into one of the tiny holes but could see nothing. He slipped the shell into his pocket. It was a zoological encounter that would preoccupy him for much of the next twenty years of his life.

Back on the *Beagle* he placed the shell under his microscope and peered down into one of the holes, adjusting his lens. There *was* a tiny creature at the base of the hole with a soft, cream-coloured body. Cemented into the hole by its head, it was upside down, waving six pairs of jointed legs in the air. Darwin was hesitant; this 'ill-formed monster' looked like an acorn barnacle, he reflected. But how could that be? The one accepted common feature of the acorn barnacle was that it built a house for itself, a cone-shaped house glued to the rocks. If it was a barnacle – and, looking more closely,

Darwin was more and more convinced – it was unclassified, up until this point: 'undiscovered'. Darwin jotted down a few thoughts in his notebook but hesitated about what to call this unnamed monster. This creature had personality and a past worth investigating.

Darwin will carry this Chilean barnacle on a journey around the world, from the South American beach back to London, preserved in a jar of wine spirits. When he has finished finding homes for all the 1,529 species he has collected and preserved in spirits on the *Beagle*, he will return to the puzzle that the creature's strange anatomy presents; and then he will write this Chilean barnacle's evolutionary biography – a puzzle that will take him eight years to think through.

Days later the barnacle-shaped volcano in the distance began to erupt. It reminded FitzRoy and Darwin of the brick glasshouses on Leith shore. FitzRoy described it in his narrative:

In the night, or rather from two to three the following morning, Osorno was observed in eruption, throwing up brilliant jets of flame or ignited matter, high into the darkness, while lava flowed down its steep sides in torrents, which from our distance (seventy-three miles) looked merely like red lines. . . . The apex of this cone being very acute, and the cone itself regularly formed, it bears a resemblance to a gigantic glass-house; which similitude is increased not a little by the column of smoke so frequently seen ascending.[49]

The volcano was a warning of the earthquake to follow a few days later. Darwin, in a Chilean forest flanking the sea with Syms Covington, his assistant, felt the earth tremble and a breeze move through the trees. The sea surged high on the beach. A few miles away the town Concepción was devastated. A tidal wave swept through the town, shedding wreckage everywhere. The ground opened up, rocks shattered and splintered, the Cathedral collapsed. Darwin was sympathetic to the inhabitants who had lost their relatives or their homes, but fascinated by the geological phenomena opening up in front of him. Nothing on the *Beagle* voyage had been as interesting as this, he claimed. The earthquake had raised the beach ten feet above high-water mark, leaving mussels and shellfish drowned in the air.

On the *Beagle* voyage Darwin spent a good deal of time inland, up mountains, climbing into hanging valleys, surveying extraordinary geological landscapes, speculating. Yet every where he went he found shells and fossil sea creatures in the rocks, sometimes hundreds of miles inland and sometimes on the tops of snow-capped mountains. He found oyster beds, like those deep under the waters of the Firth of Forth, on the sides of mountains. On 4 February 1835 he rode inland in southern Chile to see an oyster bed, out of which large forest trees were growing, at an elevation of 350 feet. Now he knew how such raised beaches had been created. He had seen it happen.

In moments, violent earthquakes could create raised beaches, erupting volcanoes, mountains. The formation of the Earth's crust could be dramatic and instantaneous, or infinitesimally subtle and gradual. Off the coast of Australia Darwin studied architects of the land so small that they had to be studied with a microscope, but so powerful that over millions of years they had created reef barriers that could resist the batterings of the fiercest storm waves: corals – island architects.

These low, insignificant coral islets stand and are victorious: for here another power, as antagonist to the former, takes part in the contest. The organic forces separate the atoms of carbonate of lime one by one from the foaming breakers, and unite them into a symmetrical structure. Let the hurricane tear up its thousand huge fragments; yet what will this tell against the accumulated labour of myriads of architects at work night and day, month after month? Thus do we see the soft and gelatinous body of a polypus, through the agency of vital laws, conquering the great mechanical power of the waves of an ocean, which neither the art of man, nor the inanimate works of nature could successfully resist.[50]

These tiny animals en masse could light up the sea or make islands or conquer the sea. Accumulated labour and immense time. With the corals Darwin's zoological and geological puzzles merged into one beautiful, branched, ancient enigma. Corals were animals, but they were also islands. They were individuals living in colonies, working together, independent and interdependent. They were animals that looked like plants – plants that worked like animals. He wrote to his sister Catherine from Chile in July 1834: 'Amongst Animals, on principle, I have lately determined to work chiefly amongst the zoophytes or Coralls; it is an enormous branch of the organized world, very little known or arranged, and abounding with most curious yet simple forms of structures.'[51]

Darwin returned to Britain in 1836 – 'all England seems changed', he said. The Reform Bill passed in 1832 had given the vote to all landowners and had changed the political landscape in Britain for ever, giving a great deal more power to the middle classes; but the country was also on the edge of a recession and the New Poor Law, which had abolished poor relief, was creating civil unrest and widespread destitution across the country. The gap between the urban poor and the middle classes had widened significantly. London itself was in a state of metamorphosis. At night the new gas pipes meant that the town was 'magically lit by its millions of gas lights'. In street after street, sewer workings revealed the sinews of the city, like a corpse opened up on a dissecting table. Elsewhere the shells of new railway stations rose against the London skyline like the arched bones of mammoth fossils. Like Dickens describing London in *Bleak House*, published some years later, perhaps Darwin could imagine extinct monsters in this strange place: 'Implacable November weather. As much mud in the streets, as if the waters had but newly retired from the face of the earth, and it would not be wonderful to meet a Megalosaurus, forty feet long or so, waddling like an elephantine lizard up Holborn Hill.'[52]

Darwin returned with his sea plunder and megafauna fossils; 368 pages of zoology notes, nearly 200 pages of marine invertebrate notes, a diary 770 pages long; 1,529 species bottled in wine spirits; 3,907 dried specimens, including giant tortoise shells and dozens of different stuffed or skinned finches.[53] He had stories to tell – some tall stories, like those of Baron Münchausen, and others that, like a detective, he had to finish, to trace their narratives through to completion. His task was to 'Prove animals like plants; trace gradation between associated & non associated animals. – & the story will be complete'.[54] Other zoologists back home, in the wake of John Vaughan Thompson's tall stories of barnacle metamorphosis as strange as the tales from Ovid, were clamouring for storytellers who would puzzle out such wondrous accounts. Facts were indeed proving stranger than fiction.

Darwin's close friend the Revd Leonard Jenyns, who had reluctantly turned down the post of naturalist on board the *Beagle* before Darwin, was one of those passionate zoologists. In a leading article called 'The Present State of Zoology' published in 1837 in a new and influential magazine, *The Magazine of Zoology and Botany*, he called his fellows to arms. The lower animals carried the secrets of life itself, he argued; and the most enigmatic and the least understood were the marine invertebrates. Of these the barnacle, recently unmasked by John Vaughan Thompson, was the most tantalizing. Someone needed to crack the barnacle code. The acquisition of a few facts about the structure and affinities of the barnacle and the bizarre shape-shifting bodies of its marine cousins might lead to 'the most important discoveries'.[55] Zoologists should do fieldwork, spend time on one species, focus, not engage in wild speculation. Evidence was needed. Empirical research had to be done by those, such as John Vaughan Thompson and Robert Grant, who had the patience and curiosity to watch, dissect and map these marine enigmas.

There were new zoologists on the block in 1836: one man in particular Darwin wanted to know better, a young man who had studied at Edinburgh the year before him and had been working since then on the pearly nautilus, a cephalopod mollusc. This was Richard Owen, the man of the hour, the new Hunterian Professor at the Royal College of Surgeons, a tall man with great glittering eyes. He, like Robert Grant, was widely read and drew his zoological ideas from France and Germany as well as Britain; and now vital materialism was the talk of the zoology season: the idea that the vibrating motion that Darwin and others had seen in the tissues of all living matter under their microscopes was inherent in all life. German and northern European zoological theories had superseded the Parisian philosophies: the ideas of Ehrenberg, Muller, Von Baer, Rudolphi, Carus, Schwann, Tiedemann, Valentin and Treviranus. Darwin began to learn German. He sought out meetings with the influential and stimulating Owen. He needed to find homes for his specimens and intellectual allies for his ideas.

If Darwin had been afraid that he would sound, on his return, like the Baron Münchausen of naturalists, with tales that no one would believe, he soon found there were plenty of other naturalists returning from sea voyages with similar tales and speculations. From fossils to vibrating primordial granules, nature, newly unmasked by fossil hunters, geologists and microscope owners, was providing the tallest of stories all by herself.

4

Settling Down

*

And this creature, rooted to one spot through life and death, was in its infancy a free swimming animal, hovering from place to place upon delicate ciliae, till, having sown its wild oats, it settled down in life, built itself a good stone house, and became a landowner, or rather a glebae adscriptus, for ever and a day. Mysterious destiny!

Charles Kingsley, *Glaucus* (1855)

Down House, Kent, 1 October 1846 – early afternoon. Darwin, now thirty-seven years old, sits at his microscope at a table in front of a large study window looking out over a gravel drive, a low flint wall, and trees beyond – lime trees, which in October are turning gold. The main road to the village of Down[1] skirts the house on the other side of the flint wall; horses and carts pass by, drivers touching their hats respectfully to Mr Darwin framed in the window. It is a good window for microscope work, he tells visitors, for the light comes from the north. It is furnished with small tables, large, comfortable armchairs, shelves stacked with papers, bottles and books, and behind a curtain there is a small toilet.[2] He has even brought a low stool of his father's from Shrewsbury to Down, with a revolving seat and castors. Sitting on it, he can turn easily from side to side and with a push of his legs he can move quickly from desk to desk and surface to surface in the study without standing up. When he is not working, the children are allowed to use it as a boat for their games, punting themselves around the drawing room with poles. On the left side of the microscope there is a round table with radiating drawers that turn on a vertical axis. The drawers are labelled 'Best Tools', 'Rough Tools', 'Specimens', 'Preparations for Specimens', and they contain his sharpest knives, the eyepiece for his micrometer, forceps, small bottles, boxes, lengths of string, sealing wax, ink, squares of sandpaper – nothing is thrown away here. 'If you throw a thing away,' he would say to Emma, 'you are sure to want it again directly.'[3] Emma agreed. Her store cupboards had been criticized by visiting aunts for muddle and chaos, but, she would tell them, she knew where to find everything. No one need disturb Darwin unnecessarily here, for he has added a little mirror to the wall outside the study window so that he can see anyone arriving at the front door and decide whether he is 'at home'. Everything in this large study is just right, perfectly designed.

9 Down House

At the back of the large, square house the garden is tidy and recently replanted with azalea trees and flower borders, and down beyond the kitchen garden, through a door in the hedge, Darwin has just designed and built a 'thinking path', a sand walk that loops its way round the edge of a small wood. Sand and red clay lodge in the ridges of his walking boots. He walks the loop of the thinking path five times every day before lunch, and when it is wet he shelters in the summerhouse that Lewis, the local builder, has just finished building for him down at the far end of the wood. Regular walks are good for his digestion. Outside the study, across the hallway, he can hear the sound of his small children playing and the soft Scottish lilt of the nursemaid Brodie, who is preparing them for an afternoon walk. The children are six-year-old William, five-year-old Annie, two-year-old Henrietta and baby George.

He is writing a letter to Captain FitzRoy, who has just returned to England from New Zealand, where he has been governor, with his wife and children. It was a curious appointment and has not been an entirely successful one, Darwin has heard. He hasn't seen FitzRoy for many years. Both are now family men. Both have settled down. Darwin writes:

My life goes on like Clockwork, and I am fixed on the spot where I shall end it; we have four children, who & my wife are all well. My health, also, has rather improved, but I am a different man in strength and energy to what I was in old days, when I was your 'Fly-catcher' on board the Beagle; I have just finished the 3rd and last part of the Geology of the Voyage of the Beagle.[4]

'I am fixed on the spot where I shall end it.' FitzRoy too has reached the end of his free-swimming days. The larval Darwin has metamorphosed; he has found his rock. Anchored to it, he will stay here, like the adult barnacle, for the rest of his days, reproducing himself, fishing with his feet as the tide comes and goes. And his life does indeed go on like clockwork, as regular as the tides: breakfast – work – opening the morning post – work – walk – lunch – writing letters – afternoon sleep – work – rest – tea – books – bed.[5] The same every day; and this is just how he likes it: a large family house 'at the extreme verge of the world',[6] full of children, animals, specimens and routines.

Barnacle specimens are on Darwin's mind again now. The glass jar containing the minute specimens of his Chilean barnacle sits on

10 Darwin's Study

the desk in front of him; the light, refracting through the yellow
wine spirits, ripples on the ceiling behind him, casting strange
watery shapes. He is trying to decide what to do with this last and
prized *Beagle* specimen. The grand plan in 1837 was to start work
on the geology of the *Beagle* and end with the tiny invertebrates he
had collected, but this plan was sabotaged when his grant ran out.
Still, there was interesting work to do here and riddles to be worked
through. Hadn't the Revd Leonard Jenyns, his Cambridge mentor,
said all those years ago that the barnacle was one of the greatest riddles yet to solve in zoology? If you could get to understand the lowest

creatures, he said, then you had a chance of understanding the higher ones; and the barnacle was one of the lowest – so common, yet so implausible. In millions along every shoreline in the world, within their snowy cones, these hermaphrodites were fishing away with their feet, glued by their heads to rocks.

Darwin shakes the jar, peering into the yellow fluid. He will have to carve out a Latin name for this creature while he works on him. 'Arthrobalanus' he will call him for the meantime, a graft of the Latin roots of 'arthro' (jointed) to 'balanus' (barnacle); but Arthrobalanus, he remembers, just doesn't fit the things the experts have said about barnacles. *Mr* Arthrobalanus just doesn't classify. All barnacles have shells; but this barnacle – and it definitely is a barnacle – has no shell of his own. How to make sense of him? It ought to be possible – after all, classification is a simple and logical process enough, if one has enough dissection and thinking time. Didn't Jenyns always insist that a good zoologist should dedicate himself to one field, one investigation, *if he had the time*. 'It's time for empirical proof not wild surmising,' he'd said. 'Always move from observable known facts to general principles,' he'd said. The marine invertebrates are unknown territories, as strange and unmapped as Amazon forests. So he will settle to the barnacle work, stick at it, see where Mr Arthrobalanus takes him. It will help him to think through some of the current issues in comparative anatomy – issues he will need to address in his species book when he expands the essay still locked away in his drawer.

All barnacles settle down in the end. They cannot help it. Most find rocks and glue themselves down as if the slightest whim might make them long for their free-swimming existence again. Mr Arthrobalanus found a conch shell and dug out a hole inside it, a conch shell that would be washed up on a Chilean beach at a particular time in history, when a particular British naturalist was walking by – a British naturalist who would make this barnacle famous and immortalize him in a glass jar.

Earlier that summer, in 1846, Darwin had sat in the new summerhouse in the rain and congratulated himself on his good fortune. Listening to the rain drumming on the summerhouse roof and watching it drip from the leaves of the hornbeams and hazels around him, he was writing to his wife, Emma, who was also his first cousin, Emma Wedgwood. Emma had taken William, Annie and baby George to the seaside for a short holiday, leaving Charles to tend the newly planted flower garden and the two-year-old Henrietta – or 'Trotty', as she was now called, named after Toby Veck in Dickens' *The Chimes*. Emma had aunts who lived in Tenby, a fishing village in Wales, an ideal place for a holiday. She needed the rest, with four children under the age of ten; and it might have been five if, in the fraught months after they had first moved into Down House, her new baby, Mary Eleanor, had not died aged only three weeks. Now there were the house renovations to deal with and the builders, a new nursemaid and a husband who, for all his generous affability and delight in the children, was wedded to his study and his microscope.

Darwin, too, needed a holiday that summer – from his wife and the small and increasingly articulate children. The house full of the sounds of sawing and hammering, his stomach painful, he played with Trotty and then passed her over to Brodie, the nurse, so that he could take his two blue opium-based pills to settle his stomach and retire to the summerhouse for a couple of hours to write letters. The first he wrote was addressed to Emma in Tenby: 'I have been sitting in the summer-house, whilst watching the thunder-storms, & thinking what a fortunate man I am, so well off in worldly circumstances, with such dear little children, & such a Trotty, & far more than all with such a wife.'[7]

Darwin might well consider himself fortunate in 1846. These were the 'hungry forties'; famine and unemployment were increasing the divide between rich and poor across Britain and Europe. The potato crop had failed the year before and bread prices were still high because the Corn Laws, established during the Napoleonic Wars to protect the income of the landowners and wheat growers of England, placed high taxes on imported corn. Now the urban and rural poor were paying the price for this government protection of landowners' profits, particularly now that the potato crops had failed. The famine in Ireland was terrible and the newspapers were full of debates about what the government should do: intervene or not? Perhaps new crops could be introduced that would alleviate the trouble. Darwin had even looked out the Chilean potato tubers he had brought back from the Chonos Archipelago, wondering if these would provide a viable alternative to the Irish potato; but the tubes had, frustratingly, perished in storage.

Darwin had a large house, a secure income that enabled him to pursue his scientific interests undisturbed, a growing and largely healthy family, and a small estate in Lincolnshire with its own tidy income. He was all too aware that others were not so fortunate. He had seen a man crying with desperation only days before. The fore-man on the building works at Down House, John Lewis, exasperated with disagreements between the builders he had contracted in, had fired the entire team. At a time of high unemployment, the consequences for some of these men would be terrible. One builder broke down and begged Darwin to reinstate him. Darwin described the incident to Emma in his letter: 'his wife had come from a distance with a Baby & is taken very ill – The poor man was crying with misery.'[8] The Darwins persuaded Lewis to re-employ him. Employment was, for some, a matter of life or death. The Darwins, like others in their position, did what they could to help.

The Irish potato famine of 1845–6 was a terrible testimony to Thomas Malthus's theory of population growth, which Darwin had read in 1838. In his 'Essay on the Principle of Population', Malthus argued that the tendency of mankind was to reproduce at such a rate that, unless it was slowed down by natural 'checks', the number of human beings would outstrip the amount of food available to feed them. These natural checks were early death, disease, famine, epidemics and war, and they meant that generally the number of individuals existing at any one time was kept in balance with the amount of food available. This was true throughout nature, Malthus argued. War, destruction, conflict, death, famine – they were all necessary; terrible but natural and necessary. Reading this essay, Darwin had found the key to natural selection. It is the fittest who survive the conflict and the destruction, he realized: the strongest wings, the largest mouth, the longest legs. Struggle ensured adaptation of species – a further conceptual piece of the jigsaw.

Darwin was also investing in land in Lincolnshire, attaching himself more firmly to the land. His father, who would nowadays be called a venture capitalist, assured him that land was a good insurance against a stock-market crash. Property was too expensive in Kent, so Darwin decided to buy 324 acres in Lincolnshire. The land cost him £12,500 but would yield £400 a year. He would be an absentee landlord, but he was also an improving one: he set up an allotment scheme on the estate to encourage self-sufficiency; he built a new farmhouse for the tenant, and he helped to pay for the village school. When the Corn Laws were repealed in 1846, his investment proved to be less lucrative than he had expected, for the tenant farmer argued that Darwin should reduce the rent now that corn had fallen in price. Darwin, who, like many liberal intellectuals in the 1840s, had argued that the Corn Laws should be repealed – how else would the poor survive the famine? – had not expected the repeal to make such a difference to his own income. Darwin's father had not predicted this. Still, he and Emma had a joint income of £14,000 a year, which kept them more than comfortable.[9]

Darwin hadn't been sure on his return from the *Beagle* that married life would be like this. He had wanted a wife. He had envied his friends their growing families and comfortable parsonages in the country. But was this the life for him? Living in London, in Great Marlborough Street, in the late 1830s with Syms Covington, his naturalist servant, free to work late into the night, free to surround himself with the dusty clutter of specimen bottles, rocks and books, he hadn't been sure that marriage was the right thing for such a man as he. He wasn't sure that settling down might not amount to a degree of suffocation. It had taken an evening to work the problem through, with a pen and paper. Two columns – like Hamlet: 'To be or not to be?' 'Marry' and 'Not Marry' he wrote neatly across the tops of these two columns. Under 'Marry' his first word was 'Children'. His second entry was 'Constant Companion' (entered with some significant reservations and ironic hesitancies: 'Constant Companion . . . who will feel interested in one, – object to be beloved and played with. – better than a dog anyhow'). His third entry was 'Home'. Then he added, anxiously underlining the words: *'But terrible loss of time'*.

However, his other voice answered: 'My God, it is intolerable to think of spending ones whole life, like a neuter bee, working, working, & nothing after all. – No, no won't do. – Imagine living all one's day solitarily in smoky dirty London House. – Only picture to yourself a nice soft wife on a sofa with a good fire, & books & music perhaps – Compare this vision with the dingy reality of Grt. Marlboro' St.'[10]

And under 'Not Marry' there was this time problem again. If he didn't marry, 'perhaps the sentence would be degradation into indolent, idle fool'. But if he were to marry, he would have to socialize, be a family man, furnish a house: 'Eheu!! I never should know French, – or see the Continent – or go to America, or go up in a balloon, or take solitary trip to Wales – poor slave – you will be worse than a Negro.'

Then the 'Marry' voice answered again: 'Never mind my boy – Cheer up – One cannot live this solitary life, with groggy old age, friendless & cold, & childless staring one in ones face, already beginning to wrinkle. – Never mind, trust to chance – keep a sharp look out – There's many a happy slave.'[11] And here he was seven years later, in Down, sitting in a summerhouse in the rain, with a growing family and a rather more complex set of ideas about family life than he had been able to imagine in Great Marlborough Street in 1838. Children bothered him and made him weep. They charmed him over breakfast with their funny stories and irritated him when they bounced into his study, wanting string or paper for their play, or, as Annie did, to bring him the box from the hallway that contained his heavily rationed snuff, her eyes bright with mischief. They made him laugh and they loved him – sometimes they bribed him to come out of the study for a few hours to play. He liked to play – down on the drawing-room floor, tickling, down in the wood, playing hide-and-seek. They terrified him with their fragility. Little animalcules, they were to be studied and measured and watched and worried over.[12]

Darwin worried about heredity. After all, he and Emma were cousins. Within a year of marrying her, he had begun to investigate the effects of breeding and interbreeding. Then, in 1845, with Emma four months pregnant with George, three years after the death of the baby Mary, he began reading Isodore Geoffroy Saint-Hilaire's book on interbreeding and monstrosity, *Histoire Générale et Particulière des*

Anomalies. Darwin and Emma affectionately called the book *Monsters & Co*. Emma's French was considerably better than Charles's, so the two worked together on some of the more difficult passages as Emma's pregnancy progressed. Did they speak openly about their fears of inherited sickness in their children? Did these ideas alarm Emma?

The children were vulnerable, even in the protective surroundings of Down. Bessie, the new nursemaid, who was little more than a teenager herself, had taken the children and their visiting cousins for a walk in the thick frost the previous winter, carrying the baby Trotty into the woods. A young housemaid and six children all under the age of ten. The eldest boys, aged nine and eight, had run off and become separated. The boys, unable to find Bessie and the younger children, had run back to Down House to raise the alarm. Charles, calming the frightened Emma, took the butler Parslow with him. They had searched the woods as the light was failing, asking urgently at local farmhouses, calling out Bessie's name. Eventually, frightened voices had answered them from the darkening woods. Bessie was huddled in a tree with her cloak around Trotty and Annie and the two young boy cousins, aged eight and five. Darwin and Parslow had taken them to a local farmhouse for food and a warm drink and then sent them all straight to bed.

So had he been right to marry? And what of this soft wife on the sofa? Back in 1838 he had wondered if a wife would enable him to fulfil his 'destination'. Would he have more chance of becoming the butterfly he wanted to be with a wife by his side?

We poor bachelors are only half men, – creeping like caterpillars through the world, without fulfilling our destination ... Of the future I know nothing, I never look further ahead than two or three Chapters – for my life is now measured by volume, chapters & sheets & has little to do with the sun. As for a wife, that most interesting specimen in the whole series of vertebrate animals, Providence only know[s] whether I shall ever capture one or to be able to feed her if caught.[13]

Would she be luminous after dark? But Emma was far more complex than he had anticipated in his columns of logic. She was a fabulous pianist who wept during concerts. She was a fine mother who kept a calm and liberal household for their children and tolerated mess so long as the children were happily occupied. Her politics were moderate and humanitarian and she had a circle of cultured and spirited friends. She shared Darwin's interest in novels, reading to him for long hours every afternoon, not because it was her duty but because she loved books and enjoyed reading without the children interrupting every two minutes. She was a critical reader, often judgemental of the prose style and narrative decisions of the novelists she read; and though constantly pregnant she was stoical and uncomplaining, despite the fact that her own nausea almost certainly matched his own. Darwin became more and more dependent upon her as his illnesses took hold.

Down House was two hours out of London, sixteen miles from St Paul's. Darwin had found his rock forty miles from the sea, but there were ghosts of shore memories even here in rural Kent. On windy days the children said the panes of glass in the north windows tasted of salt; there was the Sand Walk flanked by trees not unlike South American shores on very hot days, and there were the water-worn pebbles he had built into the garden paths, remembered long afterwards by his granddaughter Gwen Raverat, who had often stayed at Down House as a child:

The path in front of the verandah was made of large round water-worn pebbles from some sea-beach. They were not loose but stuck down tight in moss and sand and were black and shiny, as if they had been polished. I adored those pebbles. I mean literally, adored; worshipped. . . . Long after I have forgotten all my human loves, I shall remember the smell of a gooseberry leaf or the feel of wet grass on my feet, or the pebbles in the path.[14]

And there was the chalk beneath the house – the great massy chalk downs, rocks made by the bodies of infinite numbers of infinitely old and infinitely small sea creatures. Because chalk was used in the production of wallpaper, sea creatures lined the thick walls of Down. *The Annals of Natural History* magazine encouraged its readers to look at wallpaper under the new microscopes now available, for: 'When magnified 300 diameters and penetrated with Canada balsam, [you will see] a delicate mosaic of elegant coralline animalcules, invisible to the naked eye, but, if sufficiently magnified, more beautiful than any painting that covers them.'[15]

So it was here, in October 1846, that Darwin made a decision about the next ten years of his life. He had a theory of the origin of species, a theory of mutability and common ancestry that he had been working on since 1837, when he opened his first transmutation notebook; a heretical theory, which had begun to take shape in the hold of the *Beagle* and which had found fuller shape in Darwin's London lodgings amongst the dusty bones, glass jars and rocks, the spoils of his voyage; a theory that would one day be published but which was, for the moment, shelved, postponed, 231 handwritten pages carefully copied out. It was safely filed away on his shelves and its author, fearful of his early death, had already written a set of clear instructions to his wife about how to handle its publication in the event of his death. He felt the weight and responsibility of his secret.

11   Charles Darwin and his son William, 1842

Two years earlier, in 1844, Darwin had been almost ready to release his species theory. He had even begun to try it out on his friends. He had written to the young botanist Joseph Hooker first, who, at the age of twenty-six, had just returned from four years as an assistant surgeon on board the *Erebus*, assistant to the surgeon Robert McCormick, who had been sacked from the *Beagle* in Rio. Joseph was the son of Sir William Hooker, Director of the Royal Botanic Garden at Kew;[16] he was well connected, well travelled and refreshingly open-minded about natural philosophy. When Darwin wrote to Hooker in 1844, prepared to reveal his thinking for the first time on paper, his conclusions were quite clear but he was hesitant, careful with his words, mindful that even Hooker might be shocked: 'I am almost convinced (quite contrary to the opinion I started with) that species are not (it is like confessing a murder) immutable.'[17]

By October 1844 he had been more confident and assertive, writing to the Reverend Jenyns: 'The general conclusion at which I have slowly been driven from a directly opposite conviction is that species are mutable & that allied species are co-descendants of common stocks. I know how much I open myself to reproach, for such a conclusion, but I have at least honestly and deliberately come to it.'[18]

Then, in late 1844, a small volume, bound in bright-red cloth, had appeared in bookshops around the country and sold out within a few days. A beautifully written, anonymously authored book called *Vestiges of the Natural History of Creation*. This little book was an epic, telling the story of the Earth from its beginning, spinning through time and space, documenting the beginnings of life from simple marine invertebrates up through fish, amphibians, reptiles, mammals and man. It was a sensation. Its author was Robert Chambers, an encyclopaedist, journalist and publisher from Edinburgh, but his identity would not be known for a very long time. In the winter of 1844 readers of every class, background and political and religious persuasion were talking about *Vestiges*, guessing at the author's identity, reviewing and assessing the book, discussing its implications or condemning the book as heresy.[19] 'We started out as fish?' people were asking. How ridiculous. How fascinating.[20]

Darwin, in London on 20 November 1844, had opened the book apprehensively in the noisy British Museum library, afraid that this book had superseded his own theory. What he found looked heavily botched to him and full of mistakes: 'the writing & arrangement are certainly admirable, but his geology strikes me as bad, & his zoology far worse',[21] he wrote to Hooker later. For all his mistakes, however, the author of *Vestiges* had drawn transmutation into the drawing-room conversations of Britain and had drawn out its religious and philosophical implications. Heresy, some said; scandal, others. Adam Sedgwick, Woodwardian Professor of Geology at Cambridge, called it 'A foul book [in which] gross credulity and rank infidelity joined in unlawful marriage'.[22] *Vestiges* was bad news for Darwin. He had watched the reviews and refutations continue late into 1845 and the book continue to sell in its thousands. He had been troubled, too, by the way the reviewers discredited the author of *Vestiges* for clearly not having undertaken close fieldwork. 'Mr Vestiges' was a mere speculator in science, they had said. Hooker had said as much in September 1845 and Darwin agreed with him: 'no one has the right to examine the question of species who has not minutely described many'.[23] In this climate he dared not publish.

Species are mutable. Allied species are co-descendants of common stocks. This is what Darwin's 231 secret pages amounted to – eleven simple words; a hypothesis; an idea in embryo. If this embryo was to survive, it would have to be adapted to the conditions into which it would be born. If Darwin was not to go the way of 'Mr Vestiges' or Lamarck, laughed off the scientific stage, he would have to find a way of persuading his audience that these eleven words were common sense, incontrovertible and, more than anything, *plausible. He* had to be plausible. He had to have authority – after all they had said about Mr Vestiges and his pie-in-the-sky ideas, *his* ideas had to be rooted in everyday things: common sense; empirically observable truths.

This was where Mr Arthrobalanus fitted in. Life on the planet had begun with primitive creatures such as this; but barnacles were also everywhere, along every coastline in almost every part of the world. They were at once everyday objects and one of the strangest creatures in the world; and if he could show the relationship between living barnacles and fossil barnacles thousands and perhaps millions of years old, he could show that all these varieties of barnacles had once had a common ancestor and, more importantly, he could demonstrate *why* these first primitive forms had diversified.

Mr Arthrobalanus was an orphan from a family of orphans – and an aberrant, too. In order to make sense of his deviance, Darwin had to understand what being a 'normal' barnacle meant and what the common ancestor might have looked like; and that was no easy matter. He began to send for specialist barnacle literature from specialist libraries and then he was hooked, reminded just how significant this zoological riddle was. Jenyns was right: the barnacle puzzles would not be solved in a few months' work. He would be working at this particular problem for eight long years; but he didn't know that in 1846, nor did he know at first that this extremely diverse genus would be the very epitome of mutability – a minute marine monument to mutability. He had been aware since 1837 of the need to show extreme mutability within one species; he knew how crucial it was to his theories. With Mr Arthrobalanus and his aberrant kin, he would bring down the fabric of fixed species and creationism once and for all: 'Once grant that species . . . [of] one genus may pass into each other . . . & whole fabric totters & falls. – Look abroad, study gradation study unity of type study geographical distribution study relation of fossil with recent. The fabric falls!'[24]

By October 1846 the work on the house was complete and his study newly painted. Baby George was now walking; and although neither Emma nor Charles knew it yet, some time in early October husband and wife had conceived another child, who would be born the following July – a girl child, Elizabeth. On 2 October he wrote to Hooker with his new grand plan, a schedule that would take him until 1852 if he stuck to it:

I am going to begin some papers on the lower marine animals, which will last me some months, perhaps a year, & then I shall begin looking over my ten-year-long accumulation of notes on species & varieties which, with writing, I daresay will take me five years, & then when published, I daresay I shall stand infinitely low in the opinion of all sound naturalists – so this is my prospect for the future.

Are you a good hand at inventing names: I have a quite new & curious genus of Barnacle, which I want to name, & how to invent a name completely puzzles me.[25]

The first job on this new zoological island was to map out Mr Arthrobalanus's body parts. He would need to employ an artist to do this, but he struggled to find one. Joseph Hooker offered to help, so Darwin set off with Mr Arthrobalanus in his glass jar to London on 19 October, staying with his brother Alvey in Grosvenor Square, and taking Mr Arthrobalanus on the omnibus to Kew the following day. Carrying a glass jar on an omnibus would not have been easy. The buses were usually crowded, with dirty straw on the floor and the journey through the busiest parts of London was often very slow. One traveller described the experience in the *New Monthly Magazine*: 'Here we are . . . in all six and twenty sweating citizens, jammed, crammed and squeezed into each other like so many peas in a pod . . .' There were no fixed bus stops, so passengers had to stop the bus from the roadside by banging on the roof or pulling on the horses' reins.[26]

Some time around noon Darwin reached Kew Gardens, where Hooker's father, the famous botanist, Sir William Hooker, was the Director. The two men examined the internal organs of Mr Arthobalanus under Hooker's microscope after lunch. Hooker was intrigued and began to draw. They discussed names – Hooker suggested *Cryptophialus minutus*, meaning 'minute secret bowl'. Darwin was amazed at how much more he could see with Hooker's microscope than with his own, so Hooker offered to order him a new lens from Alexander James Adie, an optician and instrument maker in Edinburgh.[27] In the meantime he would lend him a lens. The new lens, which arrived in the post from Edinburgh a week later, cost Darwin three shillings and sixpence – a wise investment that would reap him substantial rewards over the following months. Hooker agreed to continue dissecting the specimen of Mr Arthrobalanus that Darwin left with him and to finish the drawings if he could. Then the two men took a walk in the gardens of Kew and Darwin, carrying his glass jar and microscope lens, walked back to the main road to flag down an omnibus to take him back to Grosvenor Square.

This was an excellent start. Darwin was delighted with the finished drawings that the enthusiastic Hooker sent him a week later: at least twenty-one minutely sketched drawings of folds, jaws, muscles, valves, 'coffin-shaped' larvae and eggs. 'The more I read, the more singular does our little fellow appear,' he wrote enthusiastically to Hooker, enclosing the borrowed lens with thanks.[28] The two men spoke about Mr Arthrobalanus now as if he were a creature they had hatched together. They often enclosed small presents with their letters: Hooker sent chutney and glass jars and porcupine quills in November that year. Darwin sent him a sheaf of carefully worded questions about what he had seen inside Mr Arthrobalanus such as: 'Did you happen to notice, whether the cherotherium footsteps pointed upwards or to the base of the animal? I can look, if you do not remember.'[29] Barnacle conversations had begun.

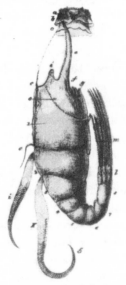

12 Mr Arthrobalanus

Throughout October Darwin dissected his specimens of Mr Arthrobalanus for several hours every day, supporting his wrists with blocks of wood.[30] The heat from the fireplace made the wine spirits evaporate and it was too cold in October to ventilate the room properly. So Darwin worked in the heady smell of wine spirits, as sweet and spicy as fine brandy. By afternoon his head was swimming and his eyes strained with the close microscope work; for Mr Arthrobalanus was no bigger than a pinhead and so Darwin could only dissect him under the microscope. Once he had extracted the body part he needed, he transferred it in a drop of water on to a glass slide. Then he would take a spoonful of gold size – a thick glue – and draw a circle around the drop of water, sealing it by fixing another glass slide to the top. To his children's eyes these glass slides, arranged in rows in the collecting drawer, were much less interesting than the jars containing shrews or lizards or octopuses curled up and floating in yellow liquid, eyes open, up on the shelves; but, for Darwin, the body parts of Mr Arthrobalanus began to be more important and frustrating than anything he had yet worked on. The more he looked, the less sense it made.

By November 1846 he had made his first major breakthrough and he wrote immediately to Hooker: 'I believe Arthrobalanus has no ovisac [ovary] at all!, & that the appearance of one is entirely owing to the splitting, & tucking up to the posterior penis, of the inner membrane of sack. – I have just found a Cirripede with an indisputable abortive anterior penis; so that this chief anomalous feature (viz two penes) in Arthrobalanus is in some degree brought within bounds.'[31]

Barnacles are hermaphrodite, he knew – they have both male and female organs; but they can't fertilize themselves. In order to fertilize their neighbours they have a very large penis, the largest penis in the animal kingdom – proportionate to size, of course. So Darwin, assuming Mr Arthrobalanus to be a hermaphrodite, like all the mapped barnacles, had searched out a penis (and found two, he thought) but couldn't find an ovisac.[32] He searched and searched, discarding specimen after specimen, *but there just wasn't one*. It didn't make sense. Lunch that day must have been frustrating. Would Darwin have explained to Emma in front of the children? Could he have kept silent, aware of the import of the missing ovisac? Had he found a pure male barnacle? Or was it that he simply hadn't searched hard enough for the ovisac that would establish its classification as a hermaphrodite? Where was it?

By November 1846, Arthrobalanus partially mapped but not yet understood, Darwin had moved on to other barnacles, searching for answers to the puzzles Mr Arthobalanus's strange body had produced. He could not understand the aberrance here, before he had come to understand the whole barnacle group; and that would take a long time. He had a theory to work with, however: Mr Arthrobalanus's aberrance had allowed him to survive, and his survival had entailed adaptation to conditions. Other living barnacles were also prime survivors. He needed to map them against their ancestors to measure their divergence from a common starting point. For fossil and living barnacle specimens he would have to fish widely. If he were to stay put in Down, he would be dependent upon the goodwill of others to collect and loan him specimens. He wrote to Richard Owen asking for any barnacle specimens he might have; he wrote to Sir James Ross, who was setting off to search for the missing Franklin expedition in the Arctic, asking him to collect any northern barnacles: 'Barnacles are so easily scraped off the rocks & put into spirits, that it would cause you but little trouble,' he reasoned.[33] He wrote to the trustees of the British Museum asking them to allow him to dissect their specimens in the privacy of his own home; and he prepared himself for months' more close dissection by ordering a new microscope. This was going to take time – longer than he had thought.

Down may have been, in Darwin's eyes, on the 'extreme verge of the world' but he had made sure when he visited the house for the first time that the village had a post office. Darwin's barnacle networks had their centre in the post office of chalky Down, one small village post office linked to the rest of the world through a reformed and increasingly efficient postal system. Rowland Hill, a British educator and tax reformer, had begun the postal reforms only in 1840, and now Darwin, like everyone else, paid a uniform rate of postage, regardless of distance. The postmaster who carried the packages up to Down House, the sound of his feet on the gravel path instantly recognizable to Darwin, carried letters and parcels with the new prepaid adhesive stamps.[34] The mail coach took the letters and parcels Darwin posted to Worcester, where they joined up with the railway network. New specially adapted railway sorting carriages were speeding up delivery times enormously.

Darwin's barnacle network would eventually stretch across Britain to several barnacle collectors in London, including Hugh Cuming, a conchologist with a major collection of fossil shells, and the trustees of British Museum collection, but also to smaller-scale collectors such as Samuel Sutchbury in Bristol, Robert Fitch, a chemist and collector of fossil barnacles in Norwich and Albany Hancock in Newcastle. The network also extended beyond British shores and around the world. Darwin wrote and received correspondence and barnacles through international postal systems: from Augustus Gould and James Dwight Dana in Boston, Alcide d'Orbigny in Paris, professors in Germany, Holland, Denmark and Norway – even from Syms Covington, his former servant, and from the father of his former shipmate, Captain King, in Australia.

A second railway boom in the mid 1840s was extending the length of the track in Britain from 2,000 to 5,000 miles. The new lines were being built by thousands of navvies, many of them Irishmen who had emigrated to England to find work during the famine.[35] Their encampments, huts made of mud or wood with tarpaulin for the roof, which accommodated as many as fifteen hundred workers and their followers, were shanty towns built by the side of railway cuttings, with their own customs, hierarchies and traditions. The men were usually regarded by local villagers as a 'race apart' with a reputation for fighting, gambling, drunkenness, disease, and theft.[36] However, survival was paramount in these communities and work was the key to survival. Most navvies worked from sixty to seventy hours a week, and work often had to be fought for. In 1846 Darwin read articles in *The Times* about riots and pitched battles amongst Irish, Scottish and English navvies at Penrith, Kendal, and Bathampton.

In a couple of decades two hundred thousand navvies achieved a feat of engineering comparable to the construction of the pyramids of Egypt. They were expected to shovel about twenty tons of earth and rock a day, excavating, cutting, banking and tunnelling. For this they were paid twenty-two shillings a week in 1846. Three per cent of the labour force were killed, fourteen per cent injured. It was these men who made postal correspondence both possible and efficient.

Railways were worth investing in for those with money to invest, as Darwin's shrewd father advised him in 1846. Darwin began to speculate, to play the stock market. His first cautious investments began to pay off. He and Emma discussed further investments, using his money and the money from Emma's trust fund. The railway boom was changing the landscape and making fortunes. The London to Brighton railway opened in 1841. Torquay was connected up to the railway network in 1848. The ten major seaside towns grew beyond all expectation in the first half of the nineteenth century as the volume of visitors encouraged entrepreneurs to build new lodging houses and the rich built their luxury summer villas. Seaside towns such as Brighton, Hastings, Margate, Torquay, Whitby and Blackpool increased in population by 254 per cent between 1801 and 1851, much higher than the rate of population growth of London during the same period.[37] Hotels and fine houses were built along the shore or close to the railway line, in long rows, overshadowing sea cottages belonging to the fishermen and women. Sea views and drawing rooms with sea air were now luxuries.

Darwin had good memories of childhood seaside visits, particularly one he made to Plas Edwards on the Welsh coast in 1819, when he was ten years old: 'The memory now flashes across me, of the pleasure I had in the evening or on a blowy day walking along the beach by myself, & seeing the gulls and cormorants wending their way home in a wild irregular course. – such poetic pleasure, felt so keenly in after years, I should not have expected so early in life.'[38] But no matter how much he wanted to replicate such rugged outdoor experiences for his own children, Darwin avoided the seaside after Edinburgh and the *Beagle* voyage. Perhaps it was the warmth and beauty of South Sea beaches that overshadowed his memories of the windy pleasures of the British coastline. Perhaps he simply didn't need to go, now that he had sea-creature collectors posted around the world. He did not need to suffer the British sea winds. Emma had always enjoyed the Tenby excursions with her children, staying with her aunt Jessie Sismondi and Harriet Surtees in South Cliff House; but as the children got older and their numbers grew, South Cliff House grew too small and, as Charles was never keen to be dragged away from Down and from his study, Emma did not press for seaside holidays. They had so much at Down.

All the bright young zoologists seemed to be settling down, Darwin reflected at the beginning of 1847. They were all looking for rocks upon which to fix themselves and reproduce. The sparkling and witty Edward Forbes, born on the Isle of Man and son of a small-time banker and timber merchant, had returned to England after following the marches of Alexander the Great across Asia Minor, discovering eighteen ancient cities and closely examining the sea creatures of the coast of Lycia, where Aristotle had walked. He had published an important and wittily illustrated book on British starfish in 1844 on his return. Now he was salaried, like Robert Grant. Edward Forbes was now Professor of Botany at King's College, London and he had been appointed to the post before he had turned thirty. He spent his summers out on the sea with increasingly large and sophisticated dredges so as to probe deeper and deeper into the dark depths for new creatures. He was even leading a campaign for dredging committees to be established across the country to share dredge building and dragging techniques amongst marine zoologists.

By 1847 Edward Forbes was in love, courting his future wife, the daughter of a general, whilst engaged day and night in writing up his new book, *A Monograph of the British Naked-Eyed Medusae* – minute luminous jellyfish, largely responsible for the phosphorescent waves Darwin had seen on the *Beagle*. Like the madrepore, like the sponge, like the sea anemone, these creatures had the most inventive ways of reproducing, only visible under a powerful microscope. They reproduced not only by producing gemmules, or eggs, but also by asexual reproduction – by *budding*. Suddenly a new creature would appear still connected to its parent, like a Siamese twin – food for thought for a man about to begin a family, Forbes reflected:

What strange and wondrous changes! Fancy an elephant with a number of little elephants sprouting from his shoulders and thighs, bunches of tusked monsters hanging epaulette-fashion from his flanks in every stage of advancement! Here a young pachyderm almost amorphous, there one more advanced, but all ears and eyes; on the right shoulder a youthful Chuny, with head, trunk, toes, no legs, and a shapeless body, on the left an infant, better grown, and struggling to get away, but his tail not sufficiently organized as yet to permit of liberty and free action! The comparison seems grotesque and absurd, but it really expresses what we have been describing as actually occurring among our naked-eyed Medusa. It is true that the latter are minute, but wonders are not the less wonderful for being packed into small compass. [39]

In fact, by late 1847 it seemed to Darwin, who now had five children (Emma had given birth to Elizabeth in July), that only Hooker was still unhooked, still unmetamorphosed; but even he was engaged: he had proposed to Henslow's eldest daughter, Frances, before he sailed for the East. Darwin wrote to his botanist friend, who was collecting plant specimens in the Himalayas, to bring him up to date on the British news: Edward Forbes, he told him was newly married – 'I almost grieve to think that I shall have no more of the old bachelor parties,' he complained. Interestingly, many of these settling naturalists were also, like him, writing long monographs about sea creatures – starfish, jellyfish, molluscs, men-of-war – trying, like him, to puzzle out taxonomies and common anatomical and reproductive patterns in the undersea world. Asexual reproduction now appeared to be more common in nature than sexual reproduction. For the women they courted, hushed parlour conversations about undersea reproduction, the slime and tentacles of marine courtship, were doubtless piquant, grotesque and erotic.

Somewhere moored out in the bay of Cape Town, in the ship the *Rattlesnake*, the twenty-two year-old Thomas Huxley, assistant ship's surgeon, was wondering about writing a monograph on molluscs but worried about the time it would take: 'I fear me that, as the old saying goes,' he wrote in his diary, 'my eyes are bigger than my belly.' It would take him at least five years, he wrote. What kind of future would marine zoology bring him? He was a man with no independent means. He would have to find a paid job on his return to England; and what about marriage – how could he possibly support a wife on his current salary?

In late 1847 the *Rattlesnake* moored in Sydney, Australia, and despite Huxley's best intentions he began to fall in love with Henrietta Heathorn, a Sydney brewer's daughter. She was captivated by the young naval surgeon's flashing eyes and fascinating manner.[40] In the ballrooms and at the dining tables of Sydney he told Henrietta not about the horrors of shipboard operations on gangrened limbs, but of the deadly beauty of the Portuguese man-of-war, the pale purple, semi-translucent *Physalis*. He had just sent a paper home to England, he explained, that would make his fortune. He was on to something. Now that he understood how all the delicate communities of polyps worked together in the *Physalis*, he said, he could *prove* they were related to jellyfish. You only had to look at the embryonic forms under a microscope to see the amazing similarities and watch the way they reproduced. Why was this paper important? Why would it make his fortune? Because it would prove that jellyfish had been wrongly classified for centuries. He would have his name on the zoological map. He would be remembered for his *Physalis* paper. And there would be more. He had zoological ground to break.

5

Better Than Castle-Building

*

I was at home
And should have been most happy – but I saw
Too far into the sea, where every maw
The greater on the less feeds evermore.
But I saw too distinct into the core
Of an eternal fierce destruction,
And so from happiness I far was gone.
Still am I sick of it; and though, today,
I've gathered young spring-leaves, and flowers gay
Of periwinkle and wild strawberry,
Still do I that most fierce destruction see –
The shark at savage prey, the hawk at pounce,
The gentle robin, like a pard or ounce,
Ravening a worm.

John Keats, Verse Letter to James Rice (April 1818)

Spring 1848, April Fool's Day. There are bluebells in the woods at Down, fringing the Sand Walk. Emma is five months pregnant; Annie, now seven, sitting in the warmth of her mother's lap, can feel the baby stir beneath the lace and brocade, beneath the taut skin of her mother's curved belly – a fluttering, she says, waiting for it to move again. Inside, in the wet darkness, ten inches long, Francis Darwin, as yet unnamed but clearly sexed, rolls and lolls freely. His eyes now open, he can see the stubby tentacles of his own fingers and the thick striped cord that floats and twists alongside him, the stalk that connects him to his mother. His mouth opens and closes, drawing in water, not air. His skin, wrinkled and translucent, stretches as his body uncurls like the flower heads of spring ferns. Emma feels his nocturnal gymnastics when her own body most needs sleep, feels him move to the sound of her voice and to the shouts of the children. She is tired, she says. She has been very sick again.

13 Emma Darwin with her son Leonard

Darwin is now fully committed to writing a full monograph on the barnacle.[1] He has given himself up to the task and will see it through to its end, however long it takes. In early 1848 he is still working on mapping the stalked barnacles; the unknown and much larger continent of the acorn, or sessile, barnacles lies on the horizon. In his study he dissects and maps another barnacle aberrant, this time from the Philippines, the *Ibla cumingii*, named after his friend, the seashell expert Hugh Cuming, who first discovered it; but this morning he is worried about the time the research is taking and about the usefulness of his growing barnacle knowledge. He hears the voices of his Cambridge mentors inside his head, disagreeing about the role of zoology. The Revd Jenyns insists that intricate fieldwork like this is the only important work in zoology,[2] yet Henslow argues that zoology should be useful: 'however delightful any scientific pursuit may be [. . .] if it shall be wholly unapplied it is of no more use than building castles in the air'.[3] Darwin feels the need to write to Henslow now, to justify these long hours and frustrations, to justify them to himself, too. So he writes: 'I fear the study of the cirripedia will ever remain "wholly unapplied" & yet I feel that such study is better than castle-building.'

For Darwin the barnacle research would always need justifying to himself and to others – after all, it was the epitome of empirical observation, field research of the most exacting kind; but so what? It could only be worth doing if there were philosophical questions that could be answered along the way. He needed to reach beyond the detail, the dry-as-dust drawing and recording and measuring of infinitely small differences within the bodies of infinitely small creatures. He needed to use the detail to explore the bigger questions, the questions about origins that now occupied his speculative mind. And he had just found something extraordinary embedded in the body of his *Ibla* specimen, something so small that it was only visible beneath a microscope, but with a significance that would take him months to think through. Certainly better than castle-building. But useful? He wasn't so sure about that.

Better than castle-building. Every day, after lunch, Darwin read *The Times* newspaper. In the spring of 1848 he read disturbing and relentless accounts of revolutions across Europe. The economic depression and famines of the 1840s were taking their toll and the poor, the unemployed and the disenfranchized were attacking castles and palaces and barricading cities in some of the most violent events in the history of Europe. In January, Sicilian citizens had risen against their Neapolitan rulers, defeating the army and declaring independence. In the last days of February a crowd of demonstrators in Paris, incensed by government corruption and abuse of power, had grown to a force strong enough to dethrone the French king and defeat the National and Municipal Guard. Darwin read *The Times* report with alarm:

France has been suddenly and violently awakened from apathy to revolution. These events in Paris will shake the kingdom with electric force. They will reverberate through Europe, where the materials of combustion are already profusely strewn abroad, and at a moment of extreme difficulty in many other countries, a sudden shock is felt from the quarter where it was least expected. These considerations disclose a most threatening and uncertain future.[4]

One hundred thousand Parisians poured on to the streets and erected hundreds of barricades all over the city using paving stones, upturned omnibuses, some four thousand trees, lamp posts and railings.[5] Louis Philippe, disguised in a pea-jacket loaned him by the English captain of the ship he sailed on, fled to England for protection.[6] Everywhere crowds sang 'The Marseillaise' as a new provisional government was assembled with poets and historians, astronomers and mechanics amongst the people's representatives, a government that established universal male suffrage, national workshops and the abolition of slavery, and reduced the working day to ten hours.[7] *The Times* had been right to predict further combustion: over the following weeks, Darwin read newspaper reports of revolutions in Hungary, the Austrian Empire, Prussia and Italy. In the great European cities – Vienna, Berlin, Milan, Venice, Frankfurt – riots toppled kings and governments and brought industries to a standstill. Struggles for power; struggles, in the hungry forties, for survival.

To many in Britain, particularly those who had read Thomas Carlyle's cataclysmic *History of the French Revolution* published in 1837, these revolutions, each seeming to spark further revolutions, seemed comparable to epic natural forces: volcanoes, earthquakes, whirlwinds and tidal waves. In March 1848, for instance, the day before Darwin penned his letter to Henslow about castles in the air, Charlotte Brontë wrote to her friend about her fears of revolution in England: 'Convulsive revolutions put back the world in all that is good, check civilisation, bring the dregs of society to its surface . . . That England may be spared the spasms, cramps and frenzy-fits now contorting the Continent, and threatening Ireland, I earnestly pray.'[8]

To many journalists, novelists and poets writing in the aftermath of these events, revolution was especially like a wave crashing against an eroding coastline or surging into cities. In *A Tale of Two Cities,* Charles Dickens described the mob as a living sea rising 'wave on wave, depth on depth' and overflowing Paris. In Arthur Hugh Clough's poem *Amours de Voyage*, written whilst Clough was in Rome during the siege of 1849, the hero, also trapped in Rome during an uprising, compares himself to a limpet clinging to a rock while the waves buffet all around him:

> So we cling to our rocks like limpets; Ocean may bluster,
> Over and under and round us; we open our shells to imbibe our
> Nourishment, close them again, and are safe, fulfilling the purpose
> Nature intended.[9]

In Down, in April and May 1848, Darwin's forays deep into the barnacle shells allowed him to see the marks, not of violent generational overthrow, but of the slowest, almost imperceptible drip-drip of time acting on shell, valve, leg, sexual organ. For all their minute gradualism, however, these changes were no less epic. While democracy struggled to establish itself against the feudal vestiges of an old order, while Darwin's friends, relations, correspondents and the journalists and columnists he read expressed their fears that this revolutionary democratic surge might spread even to conservative England, Darwin discovered evidence of an extraordinary mutation deep in the history of time: the emergence of a bisexual generation from a hermaphrodite one – a mutation of inconceivably slow gradations, but a shift no less epic in its implications.

Darwin found the evidence of this mutation marked within the soft body of the Philippino stalked barnacle *Ibla*, which, in the spring of 1848, he had carefully soaked and prised apart ready for dissection. It was grotesque and comic at the same time, this *Ibla* – quite different from Mr Arthrobalanus. One of the stalked rather than coned barnacles, it looked like a wrinkled, dark-brown hairy finger with a long claw at the end, the shrivelled finger of a witch perhaps, pickled in its bottle, in its yellowed wine spirits, several centimetres long (See opposite). For all the world as if it were beckoning; sinister and ridiculous. Cemented to a rock on the ocean

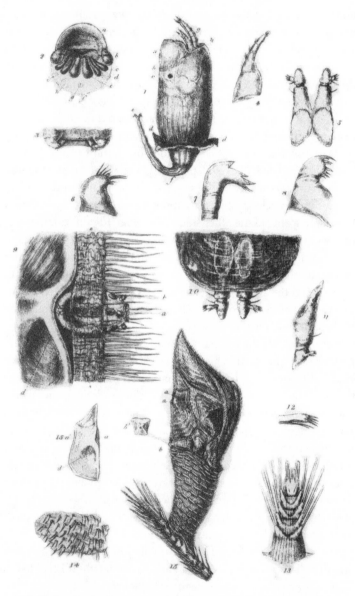

14 Ibla; drawing 9 shows the male embedded in the flesh of the female;
drawing 15 the whole specimen.

floor or secured to the bottom of a ship, these stalked barnacles moved, twisted, and oriented their bodies more than the non-stalked species, twisting in the current, directing their openings into the currents that carried food to them – fishing with their feet through the opening in their valves, just like the coned barnacles.

Being considerably larger, these *Ibla* specimens were much easier to handle than the minute Mr Arthrobalanus; but even so, barnacle dissection and identification of any kind was unbearably frustrating for Darwin.[10] Each tiny specimen had to be carefully prised from its shell, disarticulated, dissected, prepared in spirits and then mounted between two plates of glass. But as soon as he began this work, holding down the barnacle specimen to apply a knife, the barnacle would invariably spring to the other side of the room. Identification would only work through a process of scrupulous and patient elimination, building up a set of features identified from the shell, its valves and plates, the shape of the orifice at the top, the way it was attached to its rock and the type of cement the barnacle produced. All this to determine before he could even start to dissect and examine the body parts under a microscope.

Darwin, unlike the few barnacle collectors and systematists who had travelled this road before him, knew that in order to break this riddle of nature, in order to understand how barnacles had evolved the way they had, his journey must take him *inside* the horny shell – a journey into the heart of barnacle darkness, stalked or coned. In January, desperate to make the whole process easier for himself, determined to be able to see more, Darwin had invested in a new Smith and Beck microscope for £16, which made it possible to dissect under water much more easily. He wrote to describe it to Henslow: 'If you are ever starting any young naturalist with his tools, recommend him to go to Smith & Beck of 6 Colman St. City for a simple microscope: he has lately made one for me, partly from my own model & with hints from Hooker, *wonderfully* superior for coarse and fine dissections than I ever before worked with. If I had had it sooner, it would have saved me many an hour.'[11]

Whilst he could see *Ibla* with his naked eye, he had no idea of the extent of this creature's deviation from barnacle archetypes until he placed it under the new microscope. All barnacles are hermaphrodite, the zoology textbooks claimed. Hermaphroditism was what marked out barnacles from other crustacea: the barnacle hallmark. Darwin, though, had already discovered that Mr Arthrobalanus was all male and this female *Ibla*, to his astonishment, was all female, or as far as he could see – she only had female reproductive organs and no male ones. So what were the tiny creatures embedded in her flesh? More parasites? Many marine organisms ride pillion on larger organisms, living parasitically or symbiotically on their bodies. Darwin had seen thousands of tiny parasites in much of the marine flesh he had already dissected. The parasitical chain was infinite. Usually he ignored these interlopers or removed them so as not to obstruct his gaze, threw them away; but this time he didn't. This time he eased them out of the flesh of the female *Ibla* and marked them up, sandwiched within their own pair of glass slides – worth checking, he thought.

He couldn't understand what he saw under the microscope at first. The creatures living in the female *Ibla*'s flesh were, he found to his astonishment, the missing males – *Ibla* males. They were very primitive – little more than elongated tubes containing sperm (See Plate 14) – but they were very definitely male. How had this happened? Was it that food was deficient and that the creature had evolved separate sexes in order to divide up the workload of reproduction and food gathering? He couldn't answer or even speculate about this question until he had examined other closely related stalked barnacles such as the *Ibla quadrivalvis* and the *Scalpellum*.

There were further surprises. Looking again at the Australian *Ibla quadrivalvis* specimens from the British Museum collection, he found that not all of them were female; some were hermaphrodite. But both the all-female and the hermaphrodite specimens had 'complemental' males embedded in their flesh. Although bigger in relation to the females, these complemental males were exactly the same as the *Ibla cumingii* males although 'utterly different in appearance and structure' from the female and from the hermaphrodite. They seemed to represent such 'diverse beings, with scarcely anything in common, and yet all belonging to the same species'.[12] There was no other case like it in the animal kingdom, no other instance of hermaphrodites and males 'within the limits of the same species', although, Darwin noted, 'it is far from rare in the Vegetable Kingdom'.[13] Another point of connection, then, between the apparently separate kingdoms.

Darwin was sure that he had here, arranged on his desk in front of the open window, a sequence of barnacles that each represented a stage in an evolutionary sequence, like a branching tree.[14] 'Gradation', he had written in 1837; 'prove animals like plants; trace gradation between associated & non associated animals. – & the story will be complete.'[15] First he had the normal barnacles, the great majority by far, which were straightforward hermaphrodites, conforming to barnacle archetypes. Then the *Ibla* and a second, related stalked barnacle he had in his collection, *Scalpellum vulgare*, represented a transitional stage in a gradual shift towards separate sexes. These hermaphrodites still had male organs, but they were superfluous organs, now that the 'complemental' males had taken over their reproductive function. Eventually these male organs would drop away entirely or remain as much-diminished vestiges of their hermaphrodite past, like abortive stamens or pistils in flowers. Then came the completely separate-sex barnacles, representing the next stage in the evolutionary sequence: a stage in which the male was dwarfed in relation to the female and lived, as did the 'complemental' males, as nothing more than a sack of sperm. *Ibla* and *Scalpellum* were crossovers in the process of transition from hermaphroditism to separate sexes. For Darwin it provided a fascinating way of thinking about organs that were superfluous in higher animals – why men had nipples, for instance. They, too, might be vestiges of a hermaphrodite past. Curiouser and curioser; gradation after gradation.

So by 1 April, when he wrote to Henslow, he could indeed claim that this was 'better than castle-building', but he was castle-building in his own way – not castles in the air but castles built up by the bricks and mortar of chiselled fact on fact. He wasn't sure yet how he could explain the *Ibla* discovery, or what conclusions he could draw from it:

But here comes the odd fact, the male or sometimes two males, at the instant they cease being locomotive larvae become parasitic within the sack of the female, & thus fixed & half embedded in the flesh of their wives they pass their whole lives & can never move again. Is it not strange that nature should have made this one genus unisexual, & yet have fixed the males on the outside of the females . . . [16]

All this bizarre reproductive behaviour in the animal kingdom sometimes seemed to have a comic bearing on his own life, making him think about how human animals managed the reproductive process. Was it so very much superior to or more sophisticated than the barnacle way? The question was: did it work? At the end of his long April Fool's Day letter to his Cambridge mentor, Henslow, Darwin added cheerfully: 'We are all well here, & a sixth little (d) expected this summer: as for myself I have had more unwellness than usual.'[17] This was their little joke: Henslow, amazed at the speed with which Darwin and Emma were producing children, always referred to Darwin's as yet unborn children as little (d)s and Darwin, charmed, followed. Their code said something about inheritance and their shared assumptions about the male line (after all, it might have been a little (w) or a (d/w) to signal the Wedgwood inheritance). The bracketing of the baby 'd' within its protective, parenthetical shell also evoked simultaneously its as-yet-to-be-ness and its embryonic watery immersion within Emma's curved belly. Did it also remind Darwin of the soft body of the barnacle, upside down within its curved shell, fishing with its feet through the bracketed opening?

None of the barnacle discoveries provided a very serious picture of the prehistoric emergence of maleness: *thus fixed and half-embedded in the flesh of their wives they pass their whole lives and can never move again.* Nor did it provide much to support an ideology of the absolute evolutionary superiority of the male. It didn't shock Darwin, however – far from it – he found the whole barnacle sex sequence delightfully pleasing. After all, he had been thinking about all of this for some time, having worked out this theory of the gradual divergence of the sexes from an ancestral hermaphrodite in his *Notebook D* written in 1840. It had been just a theory, then – a castle in the air. He'd been searching for and recording evidence from plants to confirm his theory ever since; but this find was more significant. Here was evidence of the same sequenced branching pattern of sexual diversification in the animal world. It strengthened his commitment to wider theories about descent and modification – another piece had fallen into place, another brick of empirical evidence, not wild speculation. The evidence was all beginning to fit together.

What was newly exciting to Darwin was that his species theory was now *leading* him, directing his questions; and it was like pushing at open doors. He would never have looked closely at those specks of parasites embedded in the *Ibla* if he hadn't had a hunch, an instinct that they might provide a clue of some kind in relation to the divergence of sexes. What he had found was a polygamous animal, he wrote to Hooker a month later, in May, struggling to explain why this discovery was so important to his theory:

I never shd. have made this out, had not my species theory convinced me, that an hermaphrodite species must pass into a bisexual species by insensibly small stages, & here we have it, for the male organs in the hermaphrodite are already beginning to fail, & independent males ready formed. But I can hardly explain what I mean, & you will perhaps wish my Barnacles & Species theory al Diabolo together. But I don't care what you say, my species theory is all gospel.[18]

For some time, perhaps even since his reading of German comparative anatomists in Edinburgh, Darwin had been convinced that *embryos* – of barnacles, of humans, of lizards – were important keys to many of these zoological puzzles about form and function and evolutionary history. He was intrigued by the fact that barnacles and crustaceans repeated the same patterns of transformation in their life cycles.[19] Just comparing their adult forms seemed impossibly and blindingly limited to him by now. The clues of their origins and ancestry in deep time were evident as much in their embryological forms as in these sedentary adults.

He wasn't the first to believe so. In St Petersburg the Estonian biologist Karl Ernst von Baer, in 1848 the Professor of Comparative Anatomy and Physiology at the Medico-Surgical Academy of St

Petersburg, had been dissecting embryos for decades with similar ideas in mind. Darwin read his work enthusiastically – the questions were familiar ones. In the 1820s von Baer had shown that within an animal group – say the vertebrates – the embryos of mammals, birds, reptiles, for instance, were virtually indistinguishable from each other. The embryo literally *grew into* its reptileness or birdness – a movement away from the general form to the specific form.[20] This was how nature had worked since deep time: a process of differentiation and of separation, from the general to the particular. Von Baer also believed that there were points of comparison to be made between the embryos of higher forms and the adults of lower forms. The adult fish, for instance, resembles the human embryo. This is because the human form is more specialized than the fish, and has moved further away from the basic structural plan. Divergence is the key to adaptability. So it is not that the fish hasn't *caught up with* the human yet, rather that it has reached the point at which its own body plan fits it perfectly for its mode of existence. For von Baer, the natural world was a combination of diversity rooted in an underlying unity of form. Nature didn't travel in straight lines. It was a branching world of constant separation from a common starting point.[21]

In 1846, just as he set out on his barnacle journey, Darwin had read an important essay by the Professor of Natural History in Paris, Henri Milne-Edwards.[22] Like von Baer, Milne-Edwards believed that embryos were the keys to understanding nature's branching forms. Dissecting hundreds of crabs and lobsters in his anatomy theatres in Paris, he argued that the changes that crustaceans make in the course of their life cycle take them further and further from the *general* characteristics of the group (the embryo stage) towards their *distinctive* lobsterness or crabness. So one must look to the lobster embryo to discover the *general* characteristics of *vertebrates* and look to the adult lobster to determine the *common* features of the *species*. By looking at the path from embryo to adult, Milne-Edwards argued, it was possible to see that nature was not making straight lines from lower to higher animals but was more like 'a tree which in rising from the ground separates into several stems each of which it then divides into secondary main branches and terminates in innumerable little branches; but like the leaves with which a tree is covered, the species of animals thus produced can never, without flagrant violation of their natural relationships, be ranged in a single line'.[23]

Everyone seemed to be talking about archetypes in the 1840s and Darwin was no exception. An archetype was the basic body plan that could be seen underpinning all crustaceans or all barnacles, the plan against which their variation could be measured. So, for instance, a crustacean had twenty-one body parts, according to Milne-Edwards, and Darwin discovered that barnacles had seventeen of these. This was evidence that barnacles were a sub-group of crustaceans. Whenever he dissected a new barnacle he would check off the seventeen body parts; the mouth, for instance, would be small in some barnacles, large and elongated in others and virtually non-existent in the complemental males in which the mouth had aborted through non-use, but the seventeen body parts made up the archetype of the barnacle.

Zoologists did not agree on their interpretation of these archetypes and they used the notion of archetypes to argue for different things. For Richard Owen, these shadows and doublings and echoes between different species in adult and embryo forms could only be evidence of God, a divine architect who had designed all living forms from a single ideal pattern and he, Owen, was busy trying to discover and define that pattern. The idea that diversification had been brought about simply by progressive adaptation to environment was nonsense to him.

What could the barnacles tell Darwin about the beginnings of invertebrateness, or, for that matter, the beginnings of maleness? Darwin was finding that there were some fleshy discoveries that he could only share enthusiastically with a male friend, because their implications were both comic and chilling. The barnacle males of the *Ibla* genus, he wrote to Hooker in October, were even more grotesque and inventive than Mr Arthrobalanus. Their life cycle was entirely dedicated to reproduction: they had absolutely no other function or way of experiencing the world; at their prime they were 'mere bags of spermatozoa', in Darwin's words.[24] Once they had fertilized the giant female, they dropped away, to be replaced by other males. He wrote to Hooker, who was now in Darjeeling in the Himalayas collecting rare orchids, to tell him more about the males he had found:

I am glad to hear that you are struck with my case of the Supplemental males: I have lately reworked them most carefully. They have no mouth or stomach, but the natatory larva or rather pupa (for the larva in 2d stage in no cirripede, I find, has a mouth) fixes itself on the hermaphrodite, develops itself into a great testis! & then dies & is succeeded by a fresh crop of these temporary Supplemental males. I have caught one lately at right epoch & its entire contents were a great sperm-receptacle full of *perfect zoosperms*.[25]

Just as the older males dropped away, so they were replaced by a fresh crop of supplemental males. Nothing is indispensable: just as one creature dies, it is replaced by another; and for Darwin, visiting a dying father, this was not an easy reflection. The massive Dr Darwin, six feet four, twenty-five stone and seemingly indestructible, a rock upon which a family and whole community depended for medical and financial support, advice and investment, was dying from an illness that swelled him up and brought out boils on his body as Darwin worked on this parasitic *Ibla*. Obese and immobile, Dr Darwin couldn't turn himself over in bed and was confined to his bed or to a wheelchair, sitting for most days in the hothouse in his home in Shrewsbury, cared for by his daughters, Susan and Catherine. Darwin visited him in October and, suddenly aware of the closeness of his father's death, he wrote to the heavily pregnant Emma: 'Oh Mammy I do long to be with you & under your protection for then I feel safe.'[26]

Baby Francis was now due any day. At almost full term, his body had twisted and turned upside down in the increasingly cramped space of his mother's womb, his wrinkled feet pressing on the springy walls of the uterus, his head fixed deep in Emma's pelvis, his legs curled upwards, tucked under her ribs; he was fixed now, embedded in maternal flesh, waiting.

Darwin, as the months of the summer passed relentlessly onwards towards his father's death and his child's birth, was plagued with illness, digestive problems, 'boils & swellings' which, in the months after his father's death, would become very acute indeed. His father was dropping away – how much longer before *his* time was up?

Parslow continued to bring him his daily copy of *The Times* newspaper with its alarming reports of continuing waves of European revolution. Darwin, though concerned primarily at present with the mutants pressed between his glass microscope slides, turned his mind to politics briefly and ordered Louis Their's *History of the French Revolution* (1838) but discarded it as 'dull & poor' in comparison to Thomas Carlyle's three-volume account, which he had read some years earlier. Emma read him Currer Bell's *Jane Eyre* (1847) that summer and then Mary Wollstonecraft's *Vindication of the Rights of Women* (1792), both of which made impassioned post-revolutionary pleas for the greater independence of women.[27]

Darwin and his family, like thousands of middle-class and landed gentry across the country, had steeled themselves for the impact of the revolutionary tide, but it had been felt only as the merest ripple in Britain – some rioting in Edinburgh and in London; even the Chartist demonstrations on Kennington Common on 10 April had passed peaceably enough. The Irish people, feared in England as another source of revolution, were weakened by famine.

The British government, committed to gradual political and social reform, played a non-interventionist waiting game.[28] Iin Europe, however, the violence and revolution continued as the balance of power shifted back away from the revolutionaries. In May, Piedmont declared war on Austria, demanding independence from the Austrian Empire. In June, a hundred thousand Parisian workers, disillusioned with the ineffectiveness of the provisional government, took to the streets of Paris and were either killed, arrested or deported when government troops were ordered to suppress the uprising. In July, the Austrians defeated the Piedmontese, marched into Milan, and regained control of Vienna in October. The tide had turned.

However, there were others, like Thomas Carlyle, who saw the patterns of political revolution in terms of epic natural forces of struggle, conflict and metamorphosis. A young revolutionary in exile, Karl Marx, influenced by the writings of German philosophers Frederick Hegel and Ludvig Feuerbach, had the previous year joined a secret propaganda society called the Communist League. During the League's Second Congress in London, in late 1847, Marx and Frederick Engels had drawn up a manifesto, *The Communist Manifesto*, which had been published in February 1848. Marx and Engels described the history of political change in evolutionary and geological terms: political forms were like a creature or a land mass in a perpetual state of transformation. Just as feudalism had naturally evolved into mercantilism and then capitalism, so capitalism would inevitably give way to its most advanced successors, socialism and communism, as the necessary result of class struggle.

All previous historical movements were movements of minorities, or in the interests of minorities. The proletariat movement is the self-conscious, independent movement of the immense majority, in the interests of the immense majority. The proletariat, the lowest stratum of our present society, cannot stir, cannot raise itself up, without the whole superincumbent strata of official society being sprung into the air.[29]

Darwin and Marx were both working in very different ways to disclose the dialectical struggles inherent in the natural order, the special laws that shaped the origin, development and death of an organism and determined its replacement by a higher one. For Marx 'low' was 'high' – the rule of the proletariat, the lowest stratum of society, was the most advanced political form.[30] Progression needed conflict and struggle. Marx understood social and political change in natural and organic terms – they were like volcanoes overturning land, or earthquakes. They were dangerous, catastrophic and necessary. Darwin's evolutionary paradigms were more gradualist but no less epic.

Increasingly, Darwin was questioning the ancient natural-history divisions of so-called 'high' and 'low' organisms as he continued to develop his branching idea of nature. If nature didn't travel in straight lines but was constantly sprouting and budding off new forms from its main 'stems', then the notion of high and low forms was too restrictive and crude. This lowest and most humble of creatures, the barnacle, was none the less sophisticated, highly adapted to its environment: it was 'high' as well as 'low'. As early as 1846, on reading Henri Milne-Edwards' essay, he had written: 'Barnacles in some sense, eyes and locomotion, are lower, but then so much more complicated, that they may be considered as higher. . . . We then see that highness does not depend on perfection & number of organs, but on development. . . . leave out the term higher & lower.'[31]

That summer, on 16 August, Francis was born. This latest Darwin animalcule emerged from the mouth of Emma's belly, uncurling from its wateriness, fingernails like shells, beached and mewling and male.

In October 1848 Darwin wrote to Louis Agassiz, the Swiss naturalist now working on fossil fishes in Harvard, to give him – at Agassiz's request – a summary of his barnacle findings so far. He wrote from his father's home in Shrewsbury, for his father was now close to death. He asked the Professor to keep the contents of the letter to himself, in part because he did not want others to publish his findings and in part because he wanted to be free to alter his views and conclusions before they went to press. Darwin's letter to Agassiz is hesitant, troubled. What could he say? After all, by this point, almost exactly two years after he had begun dissecting, he ought to be able to make definitive statements about barnacles – he was probably now amongst a handful of zoologists in the world who had the authority to make such claims; but the more Darwin looked at them, the more species he had seen, sent to him from around the world, *the less he seemed to be able to say* about the group. He had plenty to say about the individuals, but he seemed no longer to be able to generalize.

Nevertheless he would try to generalize, for Agassiz's sake. After all, it was Agassiz in part who had inspired him to begin this research, for the Professor had said in 1846 in a public lecture in Boston that 'a monograph on the Cirripedia was a pressing desideratum in Zoology'.[32] He had already discovered various important and startling answers to the barnacle riddles, he wrote, but he was conscious, too, that what he had discovered was ridiculous and bizarre – 'you will laugh', he told Agassiz anxiously, 'but I assure you I would not presume to tell you anything, of which I was not sure, from repeated examinations of specimens taken at different periods & from different countries'.[33] They were definitely crustacea; there was no doubt about that now. They were 'exquisitely sensitive' to light and shadows; they could even hear and smell. Young barnacles had three pairs of legs, one compound eye and two pairs of antennae, whereas adult barnacles had six pairs of legs, no eye and one pair of antennae. One of the most remarkable parts of their natural history, he wrote, was that they attached themselves to rocks, he was sure, using a cement produced within the *ovary* of the young barnacle and secreted through its antennae.[34] This 'curious method of attachment', Darwin claimed adamantly, was the 'only character absolutely universal in the Cirripedia'. There was otherwise extreme variation wherever he looked. Every apparent universal he found was quickly demolished by the discovery of an exception. 'The great majority of cirripedes are bisexual, but it seems that they can fecundate each other, for I have scrupulously examined a Balanus, which had had its penis cut off & was imperforate, but in which the ova were impregnated.'[35]

A pregnant hermaphrodite eunuch. Sexually diverse they were – they could reproduce themselves *every which way*. Survival by diversification. Each barnacle he studied showed different deviancies from any common model he began with. What features, indeed, did all these barnacles have in common? The list was diminishing rapidly. Not even hermaphroditism was universal to barnacles, it seemed, now that Darwin had found the *Ibla*, which had separate males and females. Even the name was now a problem, for cirripede means 'curled foot', referring to the fishing feet that curl so elegantly from the hole in the tip of the cone; but some cirripedes had no feet. Darwin had discovered 'apodal' cirripedes – a contradiction in terms: 'The most remarkable individual cirripede, which I have seen is *naked*, apodal, with a suctorial mouth & parasitic in a double way within another cirripede.'[36] They were all aberrants in one way or another. Yet all were brilliantly adapted to their surroundings. Universals were dissolving the more he dissected, the more he mapped. All in the name of survival – adaptation was the name of the zoological game. Perhaps adaptation and diversification were the only universals.

A few weeks later, in November 1848, Darwin's father died and Darwin was too sick to attend his funeral. He knew his father had died an unbeliever and, if he had been a younger man, he might have been tormented by thoughts of his father's eternally damned soul. But Darwin had been drifting almost imperceptibly into the seas of unbelief for some time. He wrote later that in the 40s, 'disbelief crept over me at a very slow rate, but was at last complete. The rate was so slow that I felt no distress, and have never since doubted even for a single second that my conclusion was correct.'[37] There was no crisis, no torment of conscience and inner conflict as there had been for so many other intellectuals walking this same path away from Christian certainties, just a drift, a gradual engulfment in disbelief. By the time the letters of condolence had passed backwards and forwards in the Darwin family, it made no sense to Darwin to think of his father in heaven; but though he was not concerned about his father's soul after death, he felt his father's absence as a series of terrible physical shocks on his own body as he became more and more sick with giddiness, nausea, vomiting, boils, wakeful nights, headaches and flatulence.

A month or so after his father died, Darwin sent the three eldest children to London to have their photographs taken – then called 'daguerreotypes'. It was expensive family indulgence made more attractive to Charles and Emma by the knowledge of the inheritance Dr Darwin had left his children: Charles alone had inherited £51,000; but the death of his father had also made him more intensely aware of the relentless mutability of time, and he sought for a moment to capture his children in a still image, to take them for a moment outside time, so that later he could remember them just like this. They travelled to London with Miss Thorley to Claudet's photographic studio on the roof of the Adelaide Gallery behind St Martin's-in-the-Fields church, London. Each of the children had to sit absolutely still for fifteen seconds, unsmiling, for a smile held for so long would flicker and blur. Annie, now nearly eight, was given a basket of studio flowers to hold and, however hard she tried, she could not resist the temptation to rearrange a small white flower that had come loose in the bouquet. Her delicate long fingers are caught in movement by the daguerreotype, flurried amongst the flowers.

15 Annie Darwin in 1849

William has sharp edges and his hooded, glazed eyes show him deep in daydreams. Etty, round face framed by the half-moon of her straight bobbed hair, looks directly at us, holding herself absolutely still, hardly daring to breathe.

6

Very Like a Lobster

*

To have got the whole Barnacle family together would have been impossible for two reasons. Firstly, because no building could have held all the members and connections of that illustrious house. Secondly, because wherever there was a square yard of ground in British occupation under the sun or moon, with a public post upon it, sticking to that post was a Barnacle. No intrepid navigator could plant a flag-staff upon any spot of earth, and take possession of it in the British name, but to that spot of earth, so soon as the discovery was known, the Circumlocution Office sent out a Barnacle and a despatch-box. Thus the Barnacles were all over the world, in every direction – despatch-boxing the compass.

Charles Dickens, 'A Shoal of Barnacles', *Little Dorrit* (1855-7)

Dawn on 20 March 1849; The Lodge, Malvern, a spa town in the Malvern Hills. Darwin is sleeping alone on a single bed in a large, dark, high-ceilinged room. The door opens and a young man with long, thin hair enters carrying a pile of white towels and sheets, neatly folded and pressed. He places the towels on the chair by Darwin's bed and draws back the heavy brocade curtains. A little watery dawn light reveals the shapes of two large tin baths placed on either side of the tall windows, which look out over green hills and trees shrouded in fog. Darwin stirs, groans and wakes. The young man leaves the room while Darwin rouses himself and returns a few minutes later with two large buckets of water and various water-pouring implements, which make a good deal of noise, clattering and splashing.[1]

Darwin shakes his head nervously and retches into a small bowl next to his bed, then reaches for the small health diary that he has started to keep, with intricate notes about his daily health. The retching has diminished, he notes with resolute optimism, since he has been at Malvern. Something of all this apparent quackery *has* made a difference. The Bath Man sets about stripping the bed enthusiastically, continuing to make encouraging noises as he puts four of the white sheets, shaken out, into one of the tin baths and pours cold water over them. He has done this hundreds of times before, but it makes a change to be working in one of the lodging houses rather than in Dr Gully's establishment on the Wells Road. He has been hired for two months by Darwin from Dr Gully, the Water Cure Doctor, and now he comes every morning to The Lodge before dawn. Parslow helps to draw the morning's water.

He takes the thin mattress off the bed, places a wooden board over the metal frame and then shakes out the first of the wet white sheets, smoothing out the wrinkles. Darwin knows what to do, but he procrastinates, brushing his hair before taking off his nightgown. The man's body is tall and thin, the Bath Man notes, as he helps him to lie down on the wet sheet, reassuring him, as he does so, that the discomfort will not last long.

Darwin fixes his attention on the ceiling and concentrates on the crooning reassurances of the Bath Man, who now pulls the wet sheet across his body as quickly and tightly as he can. Darwin gasps with the cold as the Bath Man tucks the edges firmly under his recumbent body so that only his head is visible outside the wrap. Miserable, he thinks of the wrapped, embalmed bodies of the Egyptian pharaohs deep in their tombs. He thinks of swaddling clothes and Lazarus raised from the dead, confused, pinioned, wrapped around like this, dead yet alive. Best to do this quickly, the Bath Man explains, firmly, serious now – get it over with. 'You'll soon feel the benefits,' he says again. 'You'll be as warm as toast before you know it.' He places two of the other sheets across his patient and tucks them down, rolling Darwin's body over gently and smoothing out more wrinkles in the cloth. Then dry sheets and eiderdowns are piled on top, also tucked in tightly. A giant wet white chrysalis; perhaps Darwin will be a butterfly yet.

Then, with more clatters and splashes, the Bath Man is gone, leaving Darwin pinioned helplessly, damp and confined, to the bed. One and a half hours he has been prescribed for this wet-sheet wrap; one and a half hours, and it still isn't fully light.

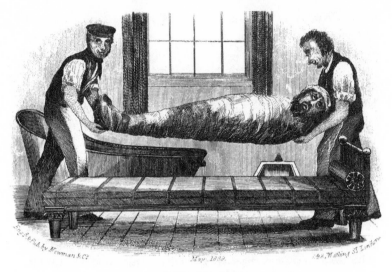

16 The Wet Sheet

Quarter past six in the morning. How preposterous. If only Hooker could see him now. He still owes his friend a letter, he remembers guiltily. He should have responded more quickly to that last letter from Darjeeling, with its delicate diagrams of the precipices and glacial spurs of the Himalayan Mountains. In it Hooker, searching for similarities in their now very different lives, explained that Darjeeling was also a kind of sanatorium. The whole ridge of Darjeeling had been purchased from the Sikkim Rajah by the East India Company in 1835 as a European refuge from the fierce heat of the Indian plains. Seven thousand feet above sea level, it was perfect for replicating the temperature, if not the climate, of London.[2] By 1849 a road had been put in place through the jungle slopes and the population had grown from 100 to 10,000 people; but political relations had always been unstable and Hooker had had to use the political agent in Darjeeling, Dr Campbell, to negotiate with the Sikkim Rajah on his behalf for permission to collect botanical specimens in his country. From Darjeeling, Hooker described heading off into the mountains, camping in the snow and in abandoned Buddhist temples, watching

the sunrise break on the mountains and sketching the shapes of pre-
cipitous valleys and gorges all around him. He was, he crowed, in
positively robust health – 'never was so well in my life' – but, he
added quickly, he would gladly share his good health with his sick
friend back home. And the sexual habits of the barnacles that
Darwin had described so excitedly in his letter of the previous
October were still in Hooker's mind, even at this altitude, though
now wrapped about with other anthropological observations: 'The
Supplemental males of the Barnacles are really wonderful, though
the supplemental males in the Bhothea families (a wife may have 10
husbands by law) have rather distracted my attention of late from
cirripedes & from our old lubrications.'[3]

Along the corridors at the other end of the house Darwin can hear
the baby, Francis, squalling and the sound of doors banging. Brodie
will be preparing the baby's breakfast. He is being weaned from his
mother's milk and he is not happy. No one else is awake yet, except
George. Darwin can hear George's characteristic and exasperating
twanging noises, made by the little jingling organ with wires stretched
inside that Emma bought him a few days before down at the toy
bazaar in the town. George, determined to see how it works, is taking it apart, piece by piece, down in the hallway somewhere.[4] Emma
will sleep longer yet – that is, if she can sleep through George's noisy
dissections. Mrs Thorley, the governess, will give the older children
their lessons and then has promised them a donkey ride today with
Emma, over the hills to St Anne's Well, the cold-water spring. Emma
and Annie in particular are happy and relaxed; they share a love for
the music here, the weekly piano and orchestral recitals and the jaunty
polkas and waltzes played in Dr Wilson's ballroom. Annie, his
beloved Annie, will want to hear all about this morning's tortures. She
will look at him again with wide-eyed incredulity.

Gradually, without his really noticing it, Darwin's body is getting
warmer. The sheets are already tepid. He has conquered the cold.
He is indeed beginning to 'feel the benefits'. Mr Gully has high
hopes for his health. Nervous dyspepsia, he diagnosed confidently,
when Darwin arrived for the first consultation in early March. The
Water Cure is strict, he explained sternly. You have to take it seri-

ously or it won't work its way. That means no sugar, salt, rich foods or stimulants. No writing, no work, minimal reading – absolutely no exertions to the brain. He would have to give up alcohol and snuff. Darwin's heart sank. He'd known this, of course, but the snuff prohibition was the worst of all.

It wasn't that he was unprepared. He'd read Gully's book when it had arrived by post from London and carefully marked it up, so he knew what to expect: 'Let no one attempt the systematic water treatment who is unwilling or unable to rid himself, for the time being, of business and botherations.'[5] But back in January he'd been desperate for anything, no matter how bizarre, to cure him of the constant pain and vomiting. He'd have given up anything – even snuff – but not the barnacles. They came to Malvern, too, he insisted, with his microscopes and notebooks. He couldn't stop now. His barnacle work had suffered from his unending sickness. But Gully refused; the barnacles were part of the problem: they were making him ill.

Darwin had been to see his cousin, the famous Dr Henry Holland in Harley Street, at some considerable expense in search of a cure for his mysterious illness, but Dr Holland had been baffled, vague in his diagnosis of 'suppressed gout'. Dr James Gully, though always circumspect, had been clear in his diagnosis and positive that he could effect at least some change; and the Water Cure was simple and harmless enough. It wasn't going to kill him. It couldn't make things any worse. After all, in February he had turned forty; something had to be done to improve his health. 'We'll be working with the curative powers of nature not artificial drugs,' Gully had said on that first day. Powerful words. Darwin had liked him from the start. They talked of Edinburgh. Though they'd not met before, they'd been contemporaries in the Medical Faculty and were almost the same age. Gully had followed Darwin's career in the medical and scientific journals, was interested in his current work and looked forward to some interesting conversations about comparative anatomy. It was some time since he had undertaken any dissections, he laughed. The poet Alfred Tennyson had been through the Water Cure most beneficially the year before, he told Darwin, and they had often spoken of the new ideas in natural philosophy. Tennyson had been particularly keen to discuss the implications of the new book *Vestiges*. He was using some of the ideas in a new poem he was writing.[6]

17 Dr Gully

Gully was unorthodox. His Water Cure sometimes included homeopathy and mesmerism as well as hydropathy. The essence of the Cure was simple, he wrote in his popular book of 1846, *The Water Cure in Chronic Disease* – it focused on making the body's natural energies heal itself. Darwin's body and mind had been abused by too much stimulation, too much thinking, Gully said when they had met in Malvern in March, and Darwin was now suffering from nervous dyspepsia. His nervous system was in complete disarray. It was no wonder that his body was complaining, retching, producing sores and swellings. It was its way of saying, 'Enough.' He'd come to the right place, Gully had said. The Water Cure would restore his body and spirit to its natural, unpolluted state in no time with vigorous exercise, daily cold douches and wrappings, plenty of Malvern water to drink and simple foods to eat: 'Nature, entirely freed from the unwholesome operation of diet, drugs, mental cares, &c, would certainly be left in the best possible position for reassuming her healthy actions.'[7]

Darwin was to leave his books behind, especially novels. Gully felt as strongly about the digestion of novels as he did about over-rich food. Four weeks of the Cure would change the depraved appetites of even the most decadent of readers. 'I have seen men', he wrote, 'whose jaded and morbid minds could previously take no nutrient save the garbage of English and French novels, devour the strong meat of History and Biography with keen and large appetite.'[8]

Gully was strict: 'The physician', he had told Darwin, 'should control, not pander to, his patient.'[9] Darwin must rest his brain if his stomach was to improve. The cold-water baths and rubbing would induce a better distribution of the blood throughout the body. For most of his patients, he explained, their troubles had started with the over-consumption of rich foods. This in its turn had caused the body's blood supply to collect around the digestive organs and became diseased. All the cold water and dry rubs would increase the circulation, draw off the bad blood and help the body regain its natural vitality and vigour.

Darwin is now feeling very warm and begins to sleep again. It is still quiet outside in the street as the sun rises. All over the small hill town, with its white stone buildings, in the bedrooms of private lodging houses, in Dr Wilson's establishment at the Crown Hotel in Abbey Road and Dr Gully's establishment at Holyrood House in Wells Road, people young and old, male and female, are being scrubbed down by Bath Attendants or wrapped up in wet sheets like Darwin. In an hour or so they will be sent out on to the hillside to walk briskly for an hour, clasping crystal-glass tumblers so that they can drink the required number of glasses of spring water from the marble fountains at the wells before breakfast, filling themselves up like water casks. The more infirm ride donkeys, and up at the tops of the hills, local girls sell bags of walnuts, biscuits, pears and ginger beer.[10] Some are very sick indeed and balanced precariously on donkeys led by servants; they look beyond recovery, though Wilson and Gully continue to fight to reinvigorate their weakened bodies from the after-effects of cholera, scarlet fever, typhoid. As the journalist Joseph Leech described in his book *Three Weeks in Wet Sheets: The Diary and Doings of a Moist Visitor to Malvern*, 'at Malvern, you met people old and yellow, and shrivelled and sapless . . . the body had become a mere shell – for the vital spark to smoulder in . . .'[11]

The Bath Man is back carrying more pails and more water. Darwin wakes rather too quickly from a deep and dreamless sleep to find himself being unwrapped, like a melon in a forcing bed. Then the whole regime begins again, just like every other morning, he writes to his sister. It's all very strange yet surprisingly pleasant:

Am scrubbed with rough towel in cold water for 2 or 3 minutes, which after the first few days, made & makes me very like a lobster – I have a washer-man, a very nice person, & he scrubs behind, whilst I scrub in front. – drink a tumbler of water & get my clothes on as quickly as possible & walk for 20 minutes . . . At the same time I put on a compress, which is a broad wet folded linen covered by mackintosh & which is 'refreshed' – ie dipt in cold water every 2 hours & I wear it all day.[12]

Neptune's Girdle: a wet compress strapped to the stomach. Almost all the patients in Malvern wore it all day, up and down the hills, drinking water at the wells, playing bridge or whist in the afternoons and at mealtimes as described by Joseph Leech: 'Anybody you meet . . . like yourself, is steaming with moisture – however gorgeous the old dowager is dressed at night, she's in reality underneath as moist as a frog – the fair young beauty is but a water-lily up to her arms in that element, and the currie-eating old Indian is hissing like an urn-iron in a full suit of swaddling clothes.'[13]

Although Dr Wilson allowed his guests to mingle freely at meal-times, Dr Gully was more strict. No stimulation of any kind, he insisted, and he segregated the male and female patients in his care into two establishments joined by a bridge, nicknamed by his patients 'the Bridge of Sighs'. It was also a matter of privacy and decency, he wrote: 'By means of the bath attendants (and the uneducated will babble) the infirmities of females are liable to become known to everybody.'[14]

Darwin, increasingly reclusive and dependent on his family, could not bear the idea of such a public-school-like establishment and rented an entire house for the duration. It was expensive, but at least he could vomit in private.[15] Here, though, at the Lodge, the routines were monastic compared even to the regimes of Down. Thomas Carlyle, who visited Dr Gully's establishment in the summer of 1851, reflected in a letter to Ralph Waldo Emerson: 'It is a strange quasi-monastic – godless and yet devotional – way of life which human creatures have here, and useful to them beyond doubt. I forsee this "water cure", under better forms, will become the Ramadhan of the overworked unbelieving English in time coming, an institution they were dreadfully in want of, this long while.'[16]

For the Darwin children this was an extraordinary and exciting time. Henrietta, who was now five years old, remembered long afterwards the exact place in the road near Down where she was told that the whole family was to move together to Malvern.[17] The clarity of that memory was a measure, she recollected, of the very quiet life they led at Down. The Malvern house and countryside were full of new adventures for the children – streams cut through the mountain sides, a fountain in the garden that trickled from marble into a large stone basin, and a garden that opened on to the mountain side – more drama than Down. Beyond The Lodge there was a town full of strange people carrying tumblers of water and standing to discuss their cures on street corners. Everywhere up there on the mountain, everywhere the children turned, there were more and more pale, smartly dressed people, carrying tumblers of water; and there were rows of charmingly doe-eyed donkeys in the main street at the Donkey Exchange, forty or fifty of them with white cotton cloths covering their side saddles, each with its name embroidered across its forehead band. Next to the Donkey Exchange was Henry Lamb's Royal Library and Bazaar, where the children would run their fingers across the books, toys and stationery and the polished white ivory of the pianofortes. William went for lessons with a clergyman who took in pupils at the Ankardine House Academy and Annie, who loved music, had dancing lessons at a Dancing Academy at Pomona House.

Meanwhile, the children continued to hear their father's growls about his privations and discomforts. In a letter to her sister-in-law Emma described overhearing little Annie telling her governess that the Water Cure 'makes Papa so angry', as if this anger were part of the treatment itself.[18] His anger, though, was very much mixed with pleasures: 'I like all this very much,' he wrote. He even

18 Douche

derived pleasures from the notorious douche bath, a thunderous deluge of mountain water that was rigged up in a series of little wooden huts in the garden. Joseph Leech, the journalist, also described the strange mixture of rapture and terror that the douche inspired:

The man pulled the string, a momentary rush, like a thunderstorm, was heard above me, and the next second the water came roaring through the pipe like a lion upon its prey, and struck me on the shoulder with a merciless bang, spinning me about like a teetotum ... like a cataract the strong column broke in foamy splinters to the ground, I felt like one who fought a great sea monster ... For a minute and a half I remained under this water spout, buffeting fiercely, until the cold column had cudgelled me as hot as a coal – aye, black and blue too; but good gracious! What a glorious luxury – a nervous but still ecstatic luxury, that made you cry out at once in terror and rapture ... I had to content myself with yelling the wild Irish war whoop of the O'Donoghues.[19]

Then there were the lamp-baths in which Darwin, like Leech, sat on a stool over a lighted lamp, wrapped in sheets. As the lamp warmed the wet sheets, Leech explained, the patient began to sweat profusely:

Suddenly, as though it could bear no more, the skin opened its pores like so many flood gates, and I ran like a shoulder of mutton before the fire, or a candle held over it. It was no moderate moisture – it was a torrent, and as it fell from my forehead on my nose like rain, it tickled me terribly, but my hands were under the blankets, and I could not help myself. My bath-man looked on me in these my first melting moments, with the eye of an artist. 'It's coming beautifully,' he said with rapture.[20]

During these long treatments, these periods of pinioned indolence, Darwin was trying to resolve the classification problems that now faced him as he prepared to pull all his notes together for the first barnacle volume. The classification of barnacles was in a 'perfect chaos', he wrote to the Professor of Anatomy and Physiology of Berlin in February. Those who had gone before him had made an absolute mess of their systems of naming. 'Literally not one species is properly defined ... The subject is heart-breaking.'[21] Sometimes they gave the newly discovered barnacle several names; sometimes they named it incorrectly. Usually they were so concerned to have their own names appended to the new genus, like a flag marking the ownership of a new island (*Ibla cumingii* for instance, after its discoverer or 'author' Hugh Cuming), that they didn't bother to get the first species name properly defined. If they had been more concerned with defining not naming, Darwin complained, his job would have been made considerably easier. He wrote to Hugh Strickland, a geologist and zoologist who had started a campaign to change the naming systems in January: 'The sooner . . . an author's name [is] buried in oblivion the better . . . A naturalist shd let his reputation rest, not on the number of the species he describes, but on the general importance of his services to Nat. Hist.'[22] Some of these men acted as if they had 'actually made the species, & it was their own property,' he raved.[23]

Although Strickland could see his arguments and accept them to some extent, many others wouldn't. Hooker, from the Himalayas, warned Darwin to be careful or he would get into 'hot water'. 'You have plenty enough to trouble you,' he said. 'Naturalists are of the genus irritabili – we have associated amongst the exceptions chiefly: but the swarm of snobs with various qualification & claims for fame & who seek fame alone is still very great & by Jove old Darwin they will be down on you like Sikhs if you do not look out.'[24] Darwin shouldn't take on the swarms of snobs, he reasoned, for the conflicts would be messy and protracted – quite ungentlemanly. However, if he were to let well alone, they were sure to find that the walls of their systems would fall simply through the sheer weight of the new species that were being described. Later that year Darwin abandoned the fight, but whilst at Malvern he remembered being 'foolish and rabid against species mongers'.[25]

Whilst Gully would not allow him to dissect barnacles at Malvern, Darwin nonetheless used his letter-writing allowance to request new specimens, so that they would arrive at Down by his return. Before he left for Malvern he wrote to Johannes Peter Müller, Professor of Anatomy and Physiology in Berlin, to beg him to post a Sicilian *Alepas minuta* specimen he owned. This was, Darwin explained in his letter, the only *Alepas* he had not yet dissected, and he promised to return all the parts after dissection. Müller's rare Sicilian *Alepas*, quarter of an inch long, was housed in the university of revolutionary Berlin, a city that had suffered sieges and riots over the previous months. Later in February he received *Scalpellum* fragments from James Bowerbank, the owner of a distillery in London and a sponge specialist, who had inadvertently acquired a good number of barnacles riding pillion on the sponges he had collected or purchased; he teased them off to post to Darwin, for they were of little value to him. From Malvern Darwin wrote to Syms Covington,

his servant and assistant from the *Beagle* voyage, who was now settled in Sydney, Australia, to bring him up to date with the Darwin
family news, but also to ask him to collect any barnacles 'that adhere
(small and large) to the coast rocks or to shells or to corals thrown
up by gales, and send them to me without cleaning the animals . . .
You will remember', he added, 'that barnacles are conical little
shells, with a sort of four-valved lid on top. There are others with
long flexible footstalks, fixed to floating objects, and sometimes cast
on shore.'[26]

Within eight days of arriving at Malvern, Darwin had his first
'crisis'. This was a measure of the success of the cure and some
people waited months for it to appear. Usually it would take the
form of boils, pimples, redness or excessive sweating, and for
Darwin it was 'an eruption' of boils all over his legs. He was delighted.
The crisis was, Joseph Leech explained to his readers, 'the
Shibboleth of Hydropathy . . . desired, looked for, wished for by
everybody' and much discussed at street corners and at dinner.
Leech, in his book *Three Weeks in Wet Sheets*, describes a conversation he overheard at dinner one night:

'Mrs –, do you know what? – I have just had my fifth crisis!'

'How fortunate!' replied the other, 'you have been Mrs –; I have been a
month in the house and have had only one, and that a poor, small,
wretched, little boil, hardly worth calling a crisis.'

I stared, not knowing what to make of it; but it was clear the announce-
ment was of the utmost interest to those around the table. Mrs – with her
five crises was manifestly the object of the utmost envy.

'Only think of five,' exclaimed the ladies.

'You have been singularly privileged, Madam,' observed a gentleman in
the civil service of India, raising his eyes from the book he was reading. In
what form Mrs –'s fifth crisis appeared she did not state, and I did not
inquire; but this I know, it was not visible.[27]

As Darwin's health improved and his trust in the Water Cure
strengthened, Gully talked to him about other healing methods he
might try. After all, there were other services at Malvern, now a centre for all kinds of experimental medical and healing treatments. It
would help to be able to see inside his stomach, like a microscope,

to see what was in there that might be causing all the sickness and pain – a blockage, perhaps, or a growth of some kind. He knew of a clairvoyant, he said, who was staying at Malvern, who could see beyond skin and bone, down into the dark, mucus-lined labyrinths of artery and colon. Darwin, confident that Gully's quackery had done him some lasting good, was not taken in by this. He was sure she was a charlatan, even if Gully wasn't. There was too much uncritical faith being placed, he thought, in the spells and tricks of clairvoyants and mesmerists all across the country; but for Gully's sake he would keep an open mind as far as possible. It wasn't as if the conventional doctors were offering any better ideas. So he agreed to see the clairvoyent on condition that he was able to test her powers for himself. He prepared himself by putting a bank note in a sealed envelope. After all, if she could see into flesh and bone, into the fleshy secret recesses of the human body, surely her vision would not be obstructed by paper, thin paper? And if she could tell him what bank note it was, she could keep it, he said.

The clairvoyant, insulted and sensing cynicism, refused to be tested. This was not a circus show, she said. She was no bearded lady performing in a tent for money. Her maid at home would perform such tricks for financial reward, she said. She had a reputation to maintain and references to prove her status, so if Darwin didn't want her to look inside him, if he wasn't a believer, why should she waste her time? Shamed and suffering a sharp look from the Water Doctor, Darwin lifted his shirt, lay back on the couch and let the clairvoyant focus her eyes down upon his prostrate stomach, for all intents and purposes like a barnacle under one of his own watch glasses. Languid, tired by the presence of an unbeliever, the clairvoyant dimmed the lights, applied herself, swaying slightly, loosing herself from the restraints of common vision, then, after a long pause, gasped and turned away. Pressed by Gully, she described what she had seen: horrors, she said – appalling horrors – a vision worse than she had even imagined possible. She would say no more. Her powers were only human, she said. There were times when it was better not to go on.[28]

If she had cared to look inside Emma Darwin's abdomen that same day, she might have seen the causes of Emma's new bout of sickness. She was, Darwin wrote to Henslow in June, in her 'usual wretched state, which to none of our friends requires any further explanation'.[29] Emma was again in the early stages of pregnancy, expecting her next baby – this would be her seventh – in January 1850. Leonard had been conceived here at Malvern in April. Darwin, reminiscing about the days before his sickness and mindful of the anatomical variety in nature, particularly in aquatic nature, joked to his cousin: 'those were delightful days when one had no such organ as a stomach, only a mouth and the masticating appurtenances'.[30] All mouth, no stomach, no bowel; painless pleasures. His stomach was his enemy now.

19 Joseph Hooker

Hooker wrote again in June, this time from Camp Sikkim in the Himalayas, where he had camped 12,000 feet up in the mountains, ten miles south of the Tibetan border. He had virtually gone native and often went days without eating, he wrote, living off wild fruits and herbs. He was in some considerable danger, too, because the Sikkim Rajah had refused to let him cross into Tibet, even though he knew that some of the rhododendron specimens Hooker needed were tantalizingly over the border, beyond his reach. The Rajah's men had assembled across the pass to prevent him going any further; conflict seemed imminent, and he had only a few of Campbell's men with him. 'I am writing on my knee on top of a great rock with a little Tent 8 ft by 6 ft over me & a blazing fire in front; still the ground is sodden and I cannot keep my feet warm', he complained. Everywhere around him up here, the rocks spoke of violence: 'the smallest fragment you can pick up shows signs of violence in the land, it is disturbed to the smallest lamination, shaken at one time, fractured at a second, baked at a third, broken away by a fifth & carried to where you find it by another. The Chaotic confusion of the jagged mass shooting up all around me must be seen to be appreciated.'[31] The enormous variation he had seen amongst the spectacular rhododendrons he was collecting were always, he said, 'asking me the vexed question, where do we come from?'

Back in Down, in July, the reassembled barnacles were asking Darwin the same vexed question: *where do we come from?* He was still collecting species from all around the world in an attempt to answer this question. Already he had been sent barnacles from the most important invertebrate collections in the world: from the British Museum, from the Cuming, Sutchbury, and Sowerby collections and from the Jardin des Plantes in Paris; but his health was still a priority and he was determined to continue the treatment in Down with Gully's advice. After all, the sickness had been cured, the flatulence had abated and Gully had promised to design him a daily regime for Down that would avoid bringing on a crisis. Darwin would be allowed to dissect for two hours a day to start with, on condition that he read nothing but newspapers. Emma, anxious that he sustain his improved health, monitored his schedules and sent the children to fetch him if he overran the agreed ration of dissection or reading time.

In the garden, near the well, Darwin employed the village carpenter, Lewis, to build a church-shaped hut to contain a tub with a platform in it and a huge cistern to contain 640 gallons of water. The carpenter's son, John, was employed to pump the cistern full every day. He recollected later: 'Mr Darwin came out and had a little dressing place and he'd get on the stage and go down, and pull the string, and all the water fell on him through a two-inch pipe. A douche they called it.'[32]

Whilst Parslow, the butler, served as Bath Man, scrubbing Darwin as his Malvern Bath Man had done, John also helped with the other regimes down in the bath hut and on the lawn, the dripping sheet and the lamp-baths followed by shallow baths, all carefully integrated into the Darwin day along with the monitored two hours of barnacle dissection, letter writing, walks and rest:

He used to get up at seven and I had to have the big bath outside the study on the lawn and Mr Darwin would come down [into his hut] and sit in a chair with a spirit lamp and all rolled round with blankets till the sweat poured off him in showers when he shook his head . . . I've heard him cry to Parslow, 'I'll be melted away if you don't hurry!' Then he'd get into the ice-cold bath in the open air.[33]

Whilst Darwin's own health had improved beyond all hope, however, the children were ill in July. Etty came down with a fever on 5 July, and then William on 11 July. William, now nine years old, did not recover as quickly as Etty and was in bed for three long weeks. Emma, now three or four months pregnant, bustled from bed to bed with remedies mixed in the kitchen from chemicals stored in her medicine cupboard.

Now that over 4,000 copies of the *Journal* of the *Beagle* travels had been sold around the world and reports of his ongoing barnacle research were being reported in the international journals, Darwin could afford to ask for specimens directly and without introductions from others; but he was still dependent on the goodwill of naturalists he had not met, if he was going to fulfil his ambition to dissect and describe every known barnacle. There were still gaps in his collection, particularly amongst the fossil barnacles. In August he wrote to two American zoologists with notable crustacea and shell collections – Augustus Gould, a Boston doctor, and James Dwight Dana, a geologist and zoologist – to ask for additional specimens. Like Bowerbank, all of these men had been collecting barnacles inadvertently as pillion passengers amongst their shell, mineral and sponge collections. Now that he had done the rounds of the sponge and shell collectors, he would have to begin approaching the geologists. There would be fossil barnacles amongst the great mineral and geology collections.

In September he and Emma travelled together to Birmingham for the British Association for the Advancement of Science meeting, in the hope that Darwin might meet other potential barnacle suppliers and friends he had lost touch with during the Malvern months; but he was solely disappointed: the place was 'large & nasty' and Henslow, Jenyns and Fox were all elsewhere. His stomach pains and vomiting returned, and he was forced to return to Malvern for a day on his journey home for further treatment and advice from the Water Doctor.

The Association meeting had other less tangible rewards for him, though: two contacts who would be important correspondents and who would lead him into new fields. The first was Albany Hancock, a Newcastle zoologist and the chair of the Tyneside Naturalists' Field Club. He had been working on molluscs since the 1840s and had a rare barnacle in his possession, *Alcippe lampas*, which he claimed excitedly in his paper was the only barnacle to bore into shells. Darwin listened to Hancock's paper, knowing that this wasn't true. After all, Darwin had Mr Arthrobalanus and other boring barnacles in his study. He introduced himself to Hancock, congratulated him on the 'admirable paper', but also warned him that he might have to qualify his claim, because there were other barnacles with burrowing powers. The two enthusiastically agreed to correspond and disagree about the burrowing of barnacles. Darwin could tell that Hancock was not just a collector; he had also read widely and was interested in philosophical questions. Had Darwin read the new paper by the curator of the Natural History Museum in Stockholm, Sven Loven? Hancock asked, sipping his glass of port. This was about a new and very curious stalked barnacle called *Alepas squalicola*, which lived as a parasite on a North Sea shark called *Squalus*. Darwin's heart sank – another *Alepas* he hadn't yet seen. He thought he had finished with the *Alepas* when Müller had sent him the rare Sicilian specimen. He asked Hancock to send him a copy of the paper and the drawing Loven had published, even though it was written in Swedish – he would have to find a translator.

The second contact Darwin made in the tea rooms of the British Association meeting in Birmingham was a Johan Georg Forchhammer, a Professor of Mineralogy and Geology at the University of Copenhagen, who was giving a paper in the Chemistry section.[34] Not only did the Copenhagen Geological Museum have barnacle fossils among its rock specimens, he replied to Darwin's discreet enquiries, but he could send them to any address that Darwin supplied. He would also give Darwin the address of his friend Japetus Steenstrup, Professor of Zoology at the University, who had his own collection of barnacles and who was particularly interested in hermaphroditism amongst the lower animals.

Darwin followed up these leads immediately on his return from Birmingham, writing to Hancock in Newcastle sending him a summary of his findings about burrowing barnacles. In return he asked Hancock for a transcript of the Loven paper and for a specimen or two of the *Alcippe lampas*, promising that he would be careful not to encroach on Hancock's territory and telling him how excited he was at the prospect of working on 'a new form of cirripede'. Hancock, generous by nature and keen to assist a fellow barnacle enthusiast, posted them almost immediately, and Darwin promised him his small collection of naked molluscs in return: there was bound to be some new genus amongst the collection, he added. Darwin also wrote to Lady Lyell, wife of Sir Charles Lyell the geologist, who was an accomplished linguist, asking her to translate the Loven paper. Lady Lyell's translation of the paper, when it arrived a few weeks later, was 'as clear as daylight to me', he wrote to Charles Lyell, and he particularly approved of her coining of the word 'leglessness'.

Now he just needed to see the *Alepas* barnacle for himself. He wrote to the elusive Sven Loven in November asking him to send him a specimen or two via Professor Forchhammer in Copenhagen, who was about to ship a number of barnacles from Steenstrup and from the Copenhagen Museum to England on his behalf. As it happened, Professor Steenstrup sent his own specimen of the *Alepas squalicola* in the Forchhammer parcel. This parcel was loaded on to the steamer *Pomona*, which sailed to St Petersburg and Riga before it arrived at Gravesend, London on 20 November. Unfortunately the barnacle package had been packed inside another parcel, which was sent to a mineral dealer in London, who did not forward the rare specimens to Darwin until the following January. Darwin was distraught: 'I write in great anxiety,' he wrote to Forchhammer, 'there was a specimen of *Alepas squalicola*, which is the cirripede of all others in the world, I wish most to dissect . . . it is a terrible loss . . . I will put an advertisement in the *Times* newspaper, & offer a large reward for recovery of Parcel.'[35] Once again, however, he found, when the specimen finally arrived, that the creature had been misdefined and misnamed. It was not an *Alepas* at all but a completely new genus, which Darwin renamed *Anelasma squalicola*.

He and Hancock continued to debate the burrowing powers of barnacles throughout the autumn and winter. He was quite sure, he told Hancock, after dissecting specimens of Mr Arthrobalanus, that none of these apparently 'boring' barnacles actually dug their own holes, but rather that the pupae crawl into holes made by other creatures in their shell or rock host, and then enlarged their holes as they grew. They had, he explained, special serrated edges to their valves and these wore away at the shell or rock and, as they blunted, were moulted and replaced by new, sharper edges. He advised Hancock to take some specimens and *see for himself*, using a compound microscope.

The gratitude and affection of Darwin's letter to Hancock masks an irritation with him for not taking the care and time to look for himself before he made large-scale claims. It further proved his point that these naturalists were sometimes so keen to see their name in print next to a new discovery that they did not do their work carefully enough. Sometimes you had to just hold fire. His species theory had taught him that.

The reproductive habits of Mr Arthrobalanus and *Ibla* were on his mind again in the autumn. He wrote to Charles Lyell:

I work now every day at the Cirripedia for 2½ hours & so get on a little but very slowly – the other day I got the curious case of a unisexual, instead of hermaphrodite cirripede, in which the female had the common cirripedial character, & in two halves of the valves of her shell had two little pockets, in *each* of which she kept a little husband; I do not know of any other case where a female invariably has two husbands. – I have one still odder fact, common to several species, namely that though they are hermaphrodite, they have small additional or as I shall call them Complemental males: one specimen itself hermaphrodite had no less than seven of these complemental males attached to it. Truly the schemes and wonders of nature are illimitable . . . [36]

The other day. It was a whole year and a half since he had made this particular discovery in the body of the *Ibla* – hardly the other day. Yet it still seemed extraordinary. These long barnacle days, punctuated only by the birth of children and the daily routines of his douches and immersions, had begun to merge into one another.

Hooker, meanwhile, was having his own crisis in the Himalayan Mountains. Darwin stumbled on an account of his friend's capture in an edition of *The Times* in November and was horrified. On 7 November Hooker and Campbell, who had joined Hooker to cross the frontier into Tibet, were seized by the Rajah's troops and held hostage for six weeks. For the Rajah, watching Hooker's movements for months, this must have been the last straw. Hooker continued to collect rhododendron seeds as he was marched south by the Rajah's men. Lord Dalhousie sent in troops to Darjeeling, warning that the rajahs 'could not play fast and loose with a British subject'. The Rajah's plan backfired. As a result of the kidnapping, Dalhousie decided to claim the southern Sikkim for the British Crown.

About the same time, Annie, then Etty and two-year-old Elizabeth came down with scarlet fever. Emma was now seven months pregnant as the three girls complained of sore throats and fever, and the characteristic red spots flushed across their necks and chests and made their tongues red and inflamed. Scarlet fever – or scarlatina, as it was then called – was responsible for nearly twenty thousand deaths in 1840 alone, many of whom were children. The girls recovered after weeks of care and quarantine, but Annie in particular took a long time to convalesce and was never as strong again.

In December Darwin was still hard at work, troubled about the children, who were still weak from the fever; by Hooker, who was still apparently a captive in Sikkim, although he had actually been released just before Christmas; by his fears about Emma's imminent labour; by the missing Scandinavian precious barnacle package, which still hadn't materialized, and troubled too by his inability to work out how the larvae of *Alcippe*, *Arthrobalanus* and *Lithotrya* actually got into the shells or rocks in which they were found. Hancock's specimens had made him less sure that they were 'cookoos' burrowing into other creatures' holes and enlarging them for their own needs. Perhaps, rather, these barnacle larvae might do their own boring by secreting some kind of acid substance instead of depending on other creatures to go in first: 'I am as much as ever in the dark, whether the larva *creeps* in, or *bores* in,' he wrote to Hancock on Boxing Day, complaining, as the year was about to turn into another, 'I begin to think I shall spend my whole life on Cirripedia, so slow is my progress working only 2 to 3 hours daily.'[37]

The mid-day douches in the garden hut were more invigorating now that the temperature of the water pumped from the well regularly dipped under 40°. The children later remembered their father's determination and groans through the coldest days, as George recorded: 'I remember well one bitter cold day with the snow covering everything waiting outside until he had finished & that he came out almost blue with cold & we trotted away at a good brisk pace over the snow to the Sandwalk.'[38]

Leonard, their new baby, arrived fifteen days into the new year and the new decade. In February he wrote to Hooker in India: 'My wife desires her kindest remembrances to you; she has lately produced our fourth boy & seventh child! – a precious lot of young beggars we are rearing. – I was very bold & administered myself, before the Doctor came, Chloroform to my wife with admirable success.'[39]

The biggest tribe of all gathered in Down House were the barnacles: over two hundred specimens crowded his study and storerooms by the time Darwin's seventh child was born. There were more arriving – almost by the day – including a rare and extensive collection of fossil specimens from Robert Fitch, a Norwich fossil collector and chemist, and the longed-for Scandinavian collection, which finally arrived in January, to Darwin's delight. However, the fossil barnacles on which he was now working took up even more time than the recent ones, for the specimens, with the fleshy part of the tissue decayed, arrived simply as single valves that crumbled and broke extremely easily; 'confound & exterminate the whole tribe,' he wrote to Hooker. 'I can see no end to my work.'[40]

7

On Speculating

*

Speculate *v.i.* **1.** pursue an inquiry, meditate, form theory or conjectural opinion (*on, upon, about* subject, the nature, cause, etc., of a thing, or abs.) **2.** Make investment, engage in commercial operation, that involves risk of loss (*has been speculating in stocks, in rubber;* esp. w. implications of rashness: *is believed to speculate a good deal*); so ~ator n. [f. L *speculari* spy out, observe (*specula* watch-tower f. *specere* look) + ATE]

Concise Oxford Dictionary

In 1850 Darwin was unlikely to know that the poet Samuel Taylor Coleridge had already used his name as a verb to describe the wild surmising of his grandfather's natural history. Erasmus Darwin, like his grandson Charles, had been a man of big ideas about the origins of life and the laws of nature. 'Darwinizing', Coleridge said, 'was all surface and no content, all shell and no nut, all bark and no wood.'[1] Erasmus and Charles Darwin would become a verb, 'Darwinize', and an adjective, 'Darwinian', and even an adverb, 'Darwinically', in the hands of later writers. All of these words carried the vestiges of a meaning that Darwin feared greatly and an accusation that had been levelled at his own grandfather: *speculation* – wild surmising; castles in the air; rashness of thought and risk of loss, even bankruptcy. Erasmus had been a speculator, some said, not with money but with ideas. The author of *Vestiges* was another such speculator.

Speculation was a family business. Darwin's father, Robert Waring Darwin, had been a practising doctor in Shrewsbury, but only a third of his income ever came from medical fees. The remainder came from dividends on securities, stocks and bonds, from rent and the interest on mortgages raised on local property. He was a canny speculator, the most important financier in the region. His money went into the local infant school, the county gaol, the infirmary, the town hall and the waterworks. He owned three quarters of Shrewsbury, the local people said, and large numbers of the local aristocracy were in hock to him. All this from so little. He had arrived in Shrewsbury in 1787, the local gossip ran, with only £20 in his pocket and immediately started buying property. With the rental income from these he had invested in canals, bridges and local manufacturing industries. When Dr Darwin died, in 1848, his estate was worth £223,759.[2]

By the late 1840s railway speculation offered quick and easy fortunes for those with capital to invest, so in 1846–7, with the Doctor's advice, Charles and Emma Darwin began to speculate on the railway companies by buying shares in the London and North West Railway, the Great Western Railway and the Monmouth Canal and Railway. In 1850 Charles made further railway investments for both Emma and himself and transferred money on the advice of his financial advisors, Salt & Sons of Shrewsbury.

The word 'speculation' carried contradictory values for men like Darwin in 1850: in the world of high finance, speculation meant risk and adventure, but also a kind of gambling, and in the pages of *The Times* governments worried about its effect on national markets and bishops about its effect on the soul.[3] Elsewhere, in the world of science, speculation was caught up in the debates about Newtonian and Baconian science, about induction and deduction and the nature of hypothesizing.[4] The word was both stirring and unsettling, then, for Darwin, and in 1850 it particularly bothered him; but it was also what he was good at. It was what his mind did, despite his best efforts to anchor it down in empirical detail. It was what he had been doing since a boy, he explained to John Lubbock, the son of the local squire, who had come to him for dissection lessons. He had tried to connect things, to see the bigger picture, on the beach at Leith, in the cabin on the *Beagle*; tried to reason from the evidence of his eyes or the evidence in his mind's eye. As if in an attic in the dark, he had felt the edge of the rim of an object, a curved shape in the darkness, and for a moment, without effort, he could see in his mind's eye the wholeness of the thing – a wheel, a tennis racket, a sieve for soil. The inch of curve under his hand made a circle in his mind or at least gave him a hypothesis with which to work.

Hypotheses, Darwin told young Lubbock again and again, had to be tested and proved by collecting facts, but you had to start with something as you began to feel your way further around the shape. You had to have an idea of what you were looking for; and that's what he was doing with his barnacles: he had a hypothesis, a species theory to work with – a theory that was only a shadowy presence for John Lubbock. Sometimes Darwin was aware that he was looking through that hypothesis at his barnacles; but most of the time he was not aware of the presence of the big idea: it was a loose framework within which to work. His grandfather had made an elegant family motto from his hypothesis: *Ex omnia conchis* (All from shells). Darwin remembered seeing the phrase underneath the family coat of arms everywhere, like an advertisement, in his grandfather's house – on the side of the family carriages and on the bookplates in the library. All from shells. Everything had come from the sea and from a single aquatic filament. Darwin hadn't understood it then – it had just seemed part of his grandfather's eccentricity; but now that he had read *Zoonomia* several times, he understood its import.

In early nineteenth-century science, not everyone agreed about the value of working from a hypothesis or about the point at which you could claim that a hypothesis had been proved. The names of Francis Bacon and Sir Isaac Newton were thrown backwards and forwards in these debates – names that carried myths and stories, sacred names. Bacon, some said, had insisted that hypotheses should only emerge *from* a weight of gradually accumulated evidence; they should not be pre-formed. Newton's famous phrase, *Hypotheses non fingo*, was much bandied about: 'I never make hypotheses.' But others pointed out that Newton and Bacon had both worked from pre-formed ideas – it was, after all, impossible to avoid developing them at some point in the scientific process. What was more important, some said, was that you should be aware of the hypotheses you were working with and prepared to jettison them if they didn't work out; but what was the role of the imagination in all of this? And was the process of discovery really as rational, objective and gradual as Bacon seemed to imply? By 1850 Darwin knew that the process of discovery was much more rugged and unpredictable, and so did others. William Whewell, author of *The Philosophy of the Inductive Science* (1840), for instance, wrote in a review of Bacon's reprinted volumes of 1857: 'The theories which make the epochs of science do not even grow gradually and regularly out of the accumulation of facts. There are moments when a spring forward is made – when a multitude of known facts acquire a new meaning … Previous to such epochs, the blind heaping up of observed facts can do little or nothing for science.'[5]

Speculation was a leap in the dark, a vision in the mind's eye.[6] The word itself came from the latin 'specere' (to look) and was kin to 'speculari' (to spy out) and 'specula' (watchtower). Sometimes it began with a flight of fancy or a daydream. And Darwin was aware that, whilst the acquisition of knowledge was cumulative and gradual, great leaps of understanding might happen in the strangest of places, in the middle of the dustiest and most laborious attention to detail, when, as if after months of seeing only the textured landscape of a piece of bark under the microscope, he could suddenly see the whole wood, as if from above – suddenly see the wood for the trees. It was astonishing, breathtaking, this shift of vision, this leap in the dark.

Darwin could remember the exact spot in the road, for instance, where, travelling in his carriage, his mind had suddenly come to understand that adaptation was a key to the diversification of species.[7] The exact spot in the road: this was a kind of road-to-Damascus revelation – it was no wonder that he used the word 'conversion' to describe his acceptance of new ideas – a revelation like an earthquake, in which everything is transformed or turned about. The mind has mountains, he knew, with crags and hanging valleys that shifted imperceptibly from day to day, but then could be heaved up in moments, like the sea beach at Concepción, leaving him breathless, like the shellfish suddenly plunged upwards into air. There would be no outward sign of such a revolution, no change in his body or face to mark the epiphany; the carriage wheels continued to turn over the potholes in the road, the beetles to scurry in the undergrowth, the spiders to build their webs.

Darwin knew how susceptible he was to pictures. Things he *saw*,

or that others made him see with their descriptions, were more per-
suasive than anything, and it made him careful, intensely aware of
the power of sight as a reasoning tool. He had listened to Robert
Grant on the shores of Leith wrapping his sea sponges around with
extraordinary theories and images of transmutation and metamorphosis. Then, on the *Beagle*, Grant's way of seeing had become
overlaid with that of Alexander von Humbolt, Darwin's favourite
travel writer and naturalist; and then, eventually, overlaid again with
the powerful new vision of Charles Lyell, whose book *Principles of
Geology* he had read on board.[8]

These writers could make him see anew. He could see the world
anew through their eyes; and although, as an older man, he was more
sceptical and cautious, it was still happening. On 5 December 1849 he
wrote to congratulate the American geologist James Dana on his new
book on volcanic geology: 'last night I ascended the peaks of Tahiti
with you, & what I saw in my short excursions was most vividly
brought before me by your descriptions ... Now that I have read you,
I believe I saw at the Galapagos, at a distance, instances of the most
curious fissures of eruption.'[9] Dana's pictures had made him think
again about the mental pictures he had brought back from the
Galapagos, shapes of mountain ranges seen only from a distance, and
remembered as a shape against clouds: I *believe* I saw, but now I don't
know. Just because Dana's pictures and ideas were vivid enough to
take him up a mountain in Tahiti in his own mind, this didn't mean,
however, that the American's conclusions were true: 'your remarks
strike me as exceedingly ingenious and novel, but they have not
converted me', he added in his letter to Dana.

Darwin was always hungry for new ideas, which made him
gullible in that first encounter with words and stories and ideas on
the written page. 'It is only after considerable reflection that I perceive the weak points,' he wrote later, when, as an old man, he was
asked to summarize his mental abilities; but at the same time, he
said, 'I am not apt to follow blindly the lead of other men. I have
steadily endeavoured to keep my mind free, so as to give up any
hypothesis, however much beloved (and I cannot resist forming
one on every subject) as soon as facts are shown to be opposed to

it.' He was no sceptic, however. He liked ideas and weighed each with equal attention to claim and evidence, though following abstract logic was sometimes a struggle for him: 'My power to follow a long and purely abstract train of thought is very limited; I should, moreover, never have succeeded with metaphysics or mathematics'.[10]

Other ideas came more slowly, through dialogue, through the hundreds of letters Darwin wrote and received every year. This was speculation in the round – collective speculation, speculation by negotiation and through exchange of information. In 1850 he read James Dana's book on geology and by letter he and Charles Lyell discussed its implications, so that Darwin continued to shape his ideas about the book, weighing it up, digesting it, in dialogue with Lyell. Sometimes the correspondence was more of a sport than a conversation, such as the series of letters he exchanged with Albany Hancock in 1850, the Newcastle naturalist he had met in 1848 at the British Association meeting. These were detailed and absorbed letters about whether certain barnacles had 'boring' powers or not. The two men slugged it out throughout the year, using not pistols or boxing gloves but increasingly detailed evidence taken from their notebooks and from the lenses of their microscopes. Perhaps they each thought that the other would concede out of sheer boredom or fatigue with the weight of evidence being aimed from each side; but neither would concede, and so the letters continued. Even on Christmas Day of 1850 Darwin sat down to write to Hancock, signing off: 'I hope I have not bored you,' a pun uncharacteristically unnoticed by Darwin. Like other scientists of his day and ours, Darwin did not discover alone, but through conversation, debate and exchange, a process made possible by a postal system, made possible by railways, made possible by investment and speculation.

In 1850 Darwin was all too conscious of the power and danger of words, too: wrenching language to tell the story he needed to tell and to do the persuading he needed it to do, like James Dana taking his reader with him up the slopes of a Tahitian mountain to see the evidence conjured up for his mind's eye. He wrote to Lyell: 'Dana is dreadfully hypothetical in many parts & often as "d–d cocked sure" as Macaulay. He writes, however, so lucidly that he is very persua-sive.'[11] In his own writing he knew that if he chose the wrong word or a clumsy inflection, the walls would fall, the picture fade, the evidence collapse. It took so much time to get it right, these words poised delicately against each other within a single sentence, pat-terned, balanced, holding the reader's trust for a moment longer, just to carry him or her, like a bridge, through to the next claim.[12] If the reader accepted this claim, would she or he accept the next? How far would readers travel with him? Would they be converted? What would it take to get them to see through his eyes?

So he threw much of his writing away: the mangled sentences, the botched phrases. Then those phrases would return like ghosts to haunt him, for he would find the scraps of his torn manuscript scattered about the schoolroom when the children used the blank backs for their exquisite watercolours and stories. One morning, visiting the empty schoolroom, Charles found red-coated soldiers and gentlemen with large heads and tiny legs, drawn and painted by George, marching across the back of a piece of manuscript.[13] As he turned it over, the staccato pencilled phrases of his barnacle classifi-cation, smudged and crossed out, brought back all too clearly the memory of an afternoon's struggle to describe barnacle cement ducts and membranes. This fabulous story about the body of a water creature and how it had come to exist over unimaginable eons of time – this story of valves and apertures and anal orifices shaping in deep time – was this so very different from the fairy tales that George and Annie inscribed on his abandoned manuscripts? It was no less wonderful. He was, after all, a Baron Münchausen amongst naturalists.

The *Ibla* and *Scalpellum* discoveries were the most wonderful of all, he knew, and how to tell this particular story? He wrote to James Dana in February 1850 that *Scalpellum* '[has] revealed to me a won-derful story, but it is too long here to tell you'.[14] In writing up his conclusions to the discoveries about the sexual peculiarities of *Ibla* and *Scalpellum*, he found it impossible to maintain the critical dis-tance needed to weigh up facts rationally and posit hypotheses. Sometimes the prose would run away with itself even here, in this the most considered and dry of texts that he would ever write; wonder would overtake him at the microscopic communities of creatures he found living together within a single sac of a single barnacle. Oh brave new world that had such creatures in it:

As I am summing up the singularity of the phenomena here presented, I will allude to the marvellous assemblage of beings seen by me within the sack of an Ibla quadrivalvis, – namely, an old and young male, both minute, worm-like, destitute of a capitulum, with a great mouth, and rudimentary thorax and limbs, attached to each other and to the hermaphrodite, which latter is utterly different in appearance and structure; secondly, the four or five, free, boat-shaped larvae, with their curious prehensile antennae, two great compound eyes, no mouth, and six natatory legs; and lastly, several hundreds of the larvae in their first stage of development, globular, with horn-shaped projections on their carapaces, minute single eyes, filiformed antennae, prosciformed mouths, and only three pairs of natatory legs, what diverse beings, with scarcely anything in common, and yet all belonging to the same species![15]

Speculation was very much on his mind in 1850, in more ways than one. It was to anchor his speculative mind in fact that he had embarked on this barnacle research, and yet the barnacle facts that he had collected from every corner of the Earth were pushing his mind into even more large-scale speculation. In March Charles Henry Lardner Woodd, then twenty-nine, a distant relative and Fellow of the Geological Society, wrote to him from Oughtershaw Hall, a manor house in a tiny hamlet in the Yorkshire Dales, with a paper and a list of pressing large-scale questions about whether masses of metamorphic rock, when heated, would crumple or bow. Darwin tried to answer but ended by transforming the young man's questions into a new set of his own and then writing in exasperation: 'how awfully complicated the Phenomenon is'. Woodd's questions reminded him of his own youth and propensity to unempirical speculation. How to reply? Should he caution Woodd against speculation of this kind or encourage him in this vein? In the end he felt he had to find a way of telling Woodd tactfully to do some careful fact-collecting, to turn his head to tangible evidence, not abstract hypothesizing:

All young geologists have a great turn for speculation; I have burned my fingers pretty sharply in that way & am now perhaps become over cautious; & feel inclined to cavil at speculation when the direct & immediate effect of a cause in question cannot be shown – How neatly you draw your diagrams; I wish you would turn your attention to real sections of the earths crust, & then speculate to your hearts content on them; I can have no doubt that speculative men, with a curb on make far the best observers. [16]

So much of what he read in 1849 seemed to be about this problem: how to conclude, how and when to hypothesize? Even the little seaside book he ordered and read in August 1850, William Harvey's *Sea-Side Book*, was still advocating a purist Baconian method in insisting that naturalists should be 'patient observers . . . contented to store up facts'; that they should abstain from 'all general views that are not warranted by the amount, either of their own knowledge, or of that of the scientific world in general'. The world was full of hasty observers, Harvey warned his readers:

Deeply informed and comprehensive intellects will discover glimpses through the haze, like the looming of distant land, where common observers can see no indications of a solution, and their 'guesses at truth', being built partly on real induction, partly on skilfully-applied analogies, often open up to us correct views of the order of nature which subsequent discoveries only confirm and strengthen. Such minds will ever be cautious in advancing theories . . . but how many hasty observers . . . ignorant of the liabilities of error, and therefore despising caution, rush forward in their course, and propose to the world their fanciful schemes as important discoveries. [17]

Harvey presented a very different version of nature from Darwin's. Harvey's seashore was divinely ordered and if he was prepared to concede that seashore creatures had transformed through time, it was, he believed, because God had transformed them. Harvey's warning to young naturalists about the dangers of rash speculation was typical of many of the scientific books that Darwin was reading. What was he, Charles Darwin? A hasty observer, or a naturalist with a deeply informed and comprehensive intellect? When did one turn into the other? Where would the barnacle research leave him?

20 Microscope

Of course, it was only four years since Joseph Hooker had pressed him to undertake this damned barnacle work, Hooker who had implied that he had to earn the respect of fellow scientists by undertaking this pursuit of classification of a single minute species. Hooker was now trekking through India and Tibet, seeing all manner of extraordinary sights and speculating with Darwin's own species theory; Hooker who now – to Darwin's considerable irritation – seemed to be bored by his barnacle reflections, frustrated that Darwin seemed only to see the dusty detail and not the bigger, speculative picture. He had grown tired of the barnacles and was hungry for news of Darwin's as yet unveiled and still to be completed species theory.

In April 1850, Hooker wrote from the Botanical Gardens in Calcutta. He was staying with the Professor of Botany there, Hugh Falconer, who was busy replanting the Gardens and draining and embanking the land. Hooker had become aware from the tone of Darwin's letter that awaited him at Calcutta that he had given offence to his friend in being less than enthusiastic about his barnacles. He wrote to apologize:

Probably I spoke too strongly about your specific work & Barnacles, but really I was in periculosis [dangerous or threatening situation] when I wrote & much harassed in mind & body. – was in short seeking & finding a very great comfort in wrapping you round with all my thoughts. I remember once dreaming that you were too prone to theoretical considerations about species & unaware of certain difficulties in your own way, which I thought a more intimate acquaintance with species practically might clear up. Hence I rejoiced at your taking up a difficult genus & in a manner the best calculated to throw light on specific characters their value &c. Since then your own theories, have possessed me, without however converting me & interested as I am in the Barnacles & felt desirous of knowing in what direction they had carried your other views.[18]

Then, as if by way of entreaty to forgive, he told Darwin of his own frustrations with Falconer who, whilst Hooker had been off in the mountains, had neglected to forward his mail for five months. Hooker had written to him from prison and from the mountain camps where he had been held hostage to beg him to forward the post, but in the end he had had to implore powerful friends to act as intermediaries. To Hooker's astonishment, when he reached Calcutta, Falconer's easy good nature and generosity of spirit made him forget all his anger: 'a more amiable fellow never lived', he enthused to Darwin. 'He had no excuse to offer & plead none – I flared up & forgave him all.'[19]

It was all a matter of time. Hooker seemed ambivalent about the barnacle work not because it was a waste of time but because it was absorbing so much time. He was impatient with Darwin, wanting him to get back to the species theory that would make his name. Hooker had never expected, when he urged him to undertake this work, that it would take him so long to complete. Darwin hadn't realized it would take so long, nor be so frustrating. Two hours of work a day was all he was allowed by Gully, which meant that he constantly had to judge how to invest his time. The motto he had used on the *Beagle* was all the more important now: *Take care of the minutes*. He had never forgotten how precious time was, and now Gully's regime meant that he was constantly looking for ways to save fifteen minutes or ten minutes of working or dissecting time, anxious not to be wasteful. The study was arranged for maximum efficiency: everything within easy reach and drawers carefully labelled, wheels on his chair. No time for rummaging. Emma was the bookkeeper of his time: wherever she was in the house, she knew exactly when Darwin had over run his work allowance and sent a child to him. He knew she was right to do so; she was only carrying out Gully's orders, guarding his health. It was for the best. He liked the way she managed him.[20]

So was the barnacle work worth it? implied Hooker in that April letter. 'What direction are they taking you?' he had asked. Darwin, always inclined to play up the labour and play down the theoretical importance of his barnacle work, replied to Hooker in June, drawing attention to the way in which speculation and systematic observation played off against each other in his barnacle work:

At last I am going to press with a small, poor first fruit of my confounded cirripedia, viz the fossil pedunculated cirripedia. You ask what effect studying species has had on my variation theories; I do not think much; I have felt some difficulties more; on the other hand I have been struck (& probably unfairly from the class) with the variability of every part in some slight degree of every species: when the same organ is *rigorously* compared in many individuals I always find some slight variability, & consequently that the diagnosis of species from minute differences is always dangerous. I had thought the same parts, of the same species more resembled than they do anyhow in Cirripedia, objects cast in the same mould. Systematic work wd be easy were it not for this confounded variation, which, however, is pleasant to me as a speculatist though odious to me as a systematist.[21]

Infinite and confounded variation these barnacles offered. No two pairs of antennae were the same. He may have begun with the notion that all barnacle legs or oviducts had been 'cast in the same mould' (signs of design, signs of a creator), but now he was working on the acorn barnacles and wondering whether there had been any common 'mould' at all. Which barnacle most closely represented the typical form, the type from which all others could be measured, when there was so much variation? This was worth four years' investment surely? For it had provided him with proof that the diagnosis of species was itself dangerous.

Whilst Darwin may have told Hooker that his first book was about to go to press in June 1850, it would not be published until the following year. There were more obstacles ahead. The problems

were caused largely by his dependence on the goodwill of fossil collectors who loaned him their collections, and on the postal system,
but also by his feeling that he had more barnacles to see before he
could write the final conclusive word. A good deal of 1850 was spent
negotiating with the collectors. Each time he needed permission
from the owner to open up a particular specimen or fossil, for
instance, he would have to write a letter and wait for its answer. The
postal system was both enabling and frustrating: parcels went astray,
letters crossed in the post, his correspondents did not answer
quickly enough. Above all, now that the first monograph was reaching completion, he had to hang on the whims of his illustrator,
James Sowerby, who was not only excruciatingly slow in his
progress with the drawings of cirripedes, but also neglected to reply
to his letters. Each month's delay meant that Darwin had to write
more letters to his collectors asking them to tolerate the loan of their
collections for another few weeks. He was embarrassed, impatient
and eventually maddened by the whole process. Then, when some
pictures did arrive from Sowerby, in April, they were inaccurate and
overly artistic, so that Darwin had to write again, this time barely able
to contain his anger and despair: 'I do not care for artistic effect, but
only for hard rigid accuracy. The inside drawings of the scuta . . . are
useless, from indistinctness and shading . . . I am sorry to . . . give
you the trouble of going over them again.' [22]

By August he began to write every week to press Sowerby:

Please remember how time slips by. – I am plagued to return specimens &
only the other day I was asked by Mr Bowerbank on part of Pal. Soc., what
progress I was making & I could only answer by stating that everything
depended on you. – All this is very disagreeable to me & I do earnestly
hope that you will endeavour to make more progress; I know not what to
say to those gentlemen, whose specimens I borrowed for only a few weeks. [23]

A week later he wrote again: 'Pray observe how time slips by,' and a
week after that: 'let me earnestly beg you to put your shoulder to the
wheel & get the job done'. His instinct for good manners and politeness was strained to its limit in these letters to Sowerby.

The book may have been drawing to a close, but Darwin still

worried about the barnacles he hadn't seen. More arrived almost by the week. There was one in Worthing he was particularly worried about; he had seen a drawing of it but felt he should dissect the actual specimen to be sure of its significance. Trouble was, the owner of the collection, Frederick Dixon, a Worthing doctor, had died the year before, and although Darwin knew where the widow lived, he couldn't bring himself to visit her. Travelling was so exhausting and took up so much time. He went to far as to write to Richard Owen, who had known Dixon, to ask him to write to his widow on his behalf and to ask for the specimen: 'I certainly do wish much to see it', he wrote to Owen, 'but not sufficiently to take me such a journey.'[24] He would never see the Worthing barnacle.

Time was indeed slipping by much faster than he was happy with. This was an anxious year. Emma had given birth to Leonard, their eighth child, in January 1850 and was pregnant again by the autumn. He wrote to Fox in October that he and Emma were thinking about emigration, another kind of life gamble: 'about midsummer we expect our 8th arrival . . . I often speculate how wise it wd be to start off to Australia, or what I fancy most, the middle States of N. America'.[25] The future of his growing tribe of boys concerned him most. William was attending a local school but would need to go to a boarding school, and Charles and Emma were not happy with the emphasis on Latin grammar that so many schools offered. They began to send away for pamphlets about progressive schools, and in the meantime Darwin continued to speculate about emigration. He wrote to Syms Covington in Australia to ask about prospects in November:

I assure you that, though I am a rich man, when I think of the future I very often ardently wish I was settled in one of our Colonies, for I have now four sons (seven children in all, and more coming), and what on earth to bring them up to I do not know. A young man may here slave for years in any profession and not make a penny. Many people think that Californian gold will half ruin all those who live on the interest of accumulated gold or capital, and if that does happen I will certainly emigrate. Whenever you write again tell me how far you think a gentleman with capital would get on in New

South Wales . . . What interest can you get for money in a safe investment?
. . . I was pleased to see the other day that you have a railway commenced,
and before they have one in any part of Italy or Turkey.[26]

Railway speculation still provided the Darwin family with a
considerable income. Since 1847 Darwin, like millions of other men
and women with capital, had been investing heavily in railways. By
the end of 1850 the paid-up share capital of UK railways amounted
to £187 million, and the scale of this investment transformed the
capital market and revolutionized the London Stock Exchange. But
by 1850, railway speculation had become more risky; the markets
were fluctuating alarmingly and dividends were down. For some
this would make a major difference to annual incomes. Charlotte
Brontë wrote, for instance, at the end of 1849:

MY DEAR SIR, – I must not thank you for, but acknowledge the receipt of
your letter. The business is certainly very bad; worse than I thought, and
much worse than my father has any idea of. In fact, the little railway property
I possessed, according to original prices, formed already a small compe-
tency for me, with my views and habits. Now, scarcely any portion of it can,
with security, be calculated upon. I must open this view of the case to my
father by degrees; and, meanwhile, wait patiently till I see how affairs are
likely to turn. . . . However the matter may terminate, I ought perhaps to be
rather thankful than dissatisfied. When I look at my own case, and compare it with that of thousands besides, I scarcely see room for a murmur.
Many, very many, are by the late strange railway system deprived almost of
their daily bread. Such then as have only lost provision laid up for the
future, should take care how they complain.[27]

The market was so gorged that shares could fall any moment and
companies made bankrupt. As with Charlotte Brontë and other
railway speculators, for Darwin it was all a question of judging the
right time to cash in the shares. A few days too early or too late and
a fortune might be lost. Charles and Emma talked anxiously about
the future of the family and its income. They were wealthy, but
Charles was not sure whether he was investing wisely enough to
ensure that there would be enough money to set up the boys. The
economic situation in 1850 seemed perilous for a man with so much
invested in the railways. All he could do was to continue to keep

meticulous accounts, watch the rise and fall of the markets and be ready when the time came to sell his shares quickly. This was where the speculation came in: being able to read the signs and gamble.

The same instinct for risk would be needed for the publication of his species theory, of course: knowing when enough facts constituted a general law, and then knowing precisely when to try out that general law on the readers who would be its judge and jury. Once that move had been made, there would be no way back.

All year Darwin had been haunted by a metaphor that Lyell had used in one of his speeches at the Geological Society. It was, Darwin wrote to him, a 'capital' metaphor. 'In a word', Lyell had said, 'the movement of the inorganic world is obvious and palpable, and might be likened to the minute-hand of a clock, the progress of which can be seen and heard, whereas the fluctuations of the living creation are nearly invisible, and resemble the motion of the hour-hand of the time-piece.'[28] It was a capital metaphor indeed, for the fluctuations of capital investments seemed to belong to the minute hand of Lyell's metaphor. The movements of the market were visible, traceable, even if they could not be predicted; but whilst the life of the children and the house and the complex financial markets in which they were all caught up were moving like the minute hand, Darwin felt himself to be disconnected, tied to the hour hand of time. Perhaps he was even moving backwards while the life of Down House hurtled forwards; certainly his health, as recorded and measured and coded in his health diaries, seemed to move backwards as well as forwards. He wrote to Lyell: 'We are all pretty flourishing here; though I have been retrograding a little, & I think I stand excitement & fatigue hardly better than in old days & this keeps me from coming to London. – My cirripedal task is an eternal one; I make no perceptible progress – I am sure that they belong to the Hour-hand, – & I groan under my task.'[29]

Behind all this urgency, the slipping away of time, was a fear that his time would run out, that his hereditary sickness would catch up with him, that he would die before he had a chance to publish his greatest speculation, the species theory. The decade had turned into another – he had not planned to be working on the barnacles in the 50s; but whilst he speculated on his own life expectancy and even felt that the Water Cure had enabled him to cheat death, buy himself a little more time, slow down the retrogression a little, he had not speculated about hereditary illness striking other parts of his family. Already, since the summer, a cluster of bacilli had begun to work its way invisibly through Annie Darwin's nine-year-old body. These bacteria and the consumption they caused were as varied and unpredictable, particularly in children, as Darwin's barnacles. They might attack anywhere, not just the lungs. A London doctor wrote in 1848 about the disease that was responsible for one in four deaths: 'Consumption, Decline or Phthisis, is the plague-spot of our climate; amongst diseases it is the most frequent and the most fatal; it is the destroying angel who claims a fourth of all who die ... Consumption steadily and surely pursues its way, and desolation of heart, of home, of hope, follow in its path.'[30] It was a disease of the hour hand; its workings were steady but imperceptible.

Annie first fell ill in the summer, when a spell of hot July weather broke in thunder and lightning. Emma noted that Annie, who was so cheerful and loving, 'never was well many days together afterwards, finding her lessons a great effort and frequently crying after she went to bed'.[31] Emma tried to find novel ways of cheering her unhappy eldest daughter, who was weak, feverish at nights, and anxious that her parents were close by at all times. In August, Emma took Annie and her governess for a carriage ride to Knole House, and a few days later she persuaded Charles to come with her to take the eldest children to stay with Uncle Joe and Aunt Caroline at Leith Hill Place, where the children collected bilberries. In October Emma sent Annie and Etty to Ramsgate with Miss Thorley, hoping that a few weeks of sea air and bathing might improve her health. Two weeks later Charles and Emma went down to join them, and Emma noted with pleasure Annie's 'bright face in meeting us at the station'. Parents and children walked together on the pier and bathed twice. Etty made doll's shoes out of seaweed on the beach and Charles, Emma noted, 'entered into daily life with a youthfulness of enjoyment which made us feel we saw more of him in a week of holiday than in a month at home'.[32]

Two days later Annie developed a fever and headache. Emma asked for a mattress to be placed on the floor of their bedroom, and another late-summer storm broke around their lodging house and out at sea. Charles had to return to Down; he had lost too much time already and he was in complex negotiations with both Sowerby and the publishers of his book, the Ray Society. He took Etty and Miss Thorley with him, leaving Emma with Annie until she was fit to travel. Emma began to steel herself for the worst. She was now sure that there was something seriously wrong with her eldest daughter – something that was not going to be blown away by sea air – and she was anxious to protect Charles as long as possible from her worst fears. As soon as she managed to get Annie back to Down on the train, she made an appointment with Dr Henry Holland and took Annie up to London for her first consultation. She returned with Annie to see him again in November.

The women of Down House – Emma, Miss Thorley (the governess), Brodie (the children's nurse) and Great Aunt Sarah, Emma's aunt – gathered protectively around the sick and uncharacteristically fretful child. Brodie had made her a little pocketbook while she had been away at the seaside, embroidered with flowers and leaves in chenille and silver thread and tied with a red-silk ribbon. Annie smiled broadly, delighted with her neat and pretty little book. She was nine when she started this first diary. Not sure at first what to record, she adopted her mother's clipped style of diary entries and her subject matter: the weather and visitors were to be recorded regularly. Annie wrote in the very last days of November 1850:

> 23 Cicely wrote to me. Went Aunt S. Rainy morn. But fine afternoon
> 24 Very rainy all day.
> 25 Cold but rather fine morn.
> 26 Thremomiter 46.
> 27 Greata wrote to me.[33]

Annie returned her cousin's letter a few days later: 'We have got a new pony. It is rather a little one. I think your donkey sounds a very nice one. I should like to see your little white guinea-fowl. Are all the little turkeys sold? On Sunday it was raining dreadfully, and the pit in the sand walk was full of water. Is your swing taken down? Ours has been taken down a long while.'[34] In her allusions to the rain and the swing being taken down, Annie shows her own child's-eye-view sense of the passing of the year. No more swing until next summer, when it would be hung up again between the yew trees. But Annie would not be there to see it.

8

Writing Annie

*

The month of April is proverbial for its fickleness, for its intermingling showers, and flitting gleams of sunshine; for all the species of weather in one day; for a clear mixture of clear and cloudy skies, greenness and nakedness, flying hail and abounding blossoms. But to a lover of Nature, it is not the less characterised by the spirit of expectation with which it imbues the mind.

William Howitt, *The Book of the Seasons* (1831)

It is a mild January afternoon in 1851; the days begin to lengthen. This morning's post has brought Darwin the final plates of the illustrations for his completed first barnacle volume from James Sowerby in London. He sits at his desk in front of the window checking the accuracy of the drawings with a magnifying glass, making the best of the thin and fading winter light. Laid out on the table in front of him are five pieces of paper, on each of which are fifty or sixty closely engraved shapes. Some look like conical shells, others like leaves, or fans, or claws. Many are delicately ridged. The drawings are of fossil stalked barnacle valves – valves that Darwin knows inside out. The tiny originals are lined up across the top of the table so that he can compare them to the engravings one last time before he returns the collections. Most of this set on his table belong to Robert Fitch, a Norwich chemist and geologist, a man whose patience has been sorely tried: Darwin intended to borrow his rare fossil collection for only a few weeks and he has now had it for a year. He has examined the valves in every light and from every angle. He can summon these shapes and ridges and patterns when his eyes are closed; and James Sowerby has made mistakes in the drawings – not many, but enough for Darwin to ask him to make five significant changes. This volume and its successors must be a perfect record.

Now that the fossil stalked barnacle book is finished Darwin must return Fitch's collection, this fruit of twenty years of collecting – perhaps the finest collection of fossil barnacles in the world – to its owner. Although it is unlikely he will ever see these ancient objects again, he is glad to let them go. This has been frustrating and at times impossible work – trying to identify and describe ancient species, working usually from a single valve that has survived imprinted into Norfolk chalk. Several valves arrived damaged and he has had to glue them back together again – embarrassing. He is determined they will survive the return journey. He puts a small piece of silver paper on to the desk and lifts one of the smallest valves onto it with a pair of tweezers. He writes its name carefully on the label that he will attach to the pill box in which it will travel through the post: *Pollicipes angelini*, one of the most ancient stalked barnacle forms, a survivor from the early Jurassic period. This is the barnacle ancestor – not the first, but certainly amongst the earliest generations of barnacles.

The stalked barnacles came first – coning was a later variation, an adaptation to shoreline conditions. The *Pollicipes* had used the valve in Darwin's hand to fish for food in seas in which ichthyosaurs and plesiosaurs hunted. Yet the dredging boats still pulled in clusters of *Pollicipes* that were almost identical. Pulled from the seabed, they looked for all the world like dinosaur body parts: glistening black stalks as wrinkled and thick as elephant skin and a claw-like valve at the end of the stalk like a bird's beak. Inside the valve the tiny shrimp-like body fished with its legs through the valve opening, its legs looking every bit like the black-dyed ostrich feathers on this season's mourning hats. No fundamental change since Jurassic times, just diversification: no need for change; perfect adaptation to environment. Yet even in Jurassic times there had been thirty-two different species of barnacle, Darwin told his readers. The fossil barnacles were a testimony to 'the exhaustless fertility of nature in the production of diversified yet constant forms'.[1] Now Nature's exhaustless prodigality has been recorded, marked down minutely. *Pollicipes*, *Scalpellum* and *Ibla* fossils from all over the world have been defined, identified, placed in family groupings; all from valves alone.

The barnacles are far from finished with Darwin, however. Now that he has completed the classification of the thirty-two fossil stalked barnacles, he must classify and describe the recent stalked barnacles, and there are hundreds of these. The smaller group of fossil acorn barnacles will make a slim volume, but the classification of the largest group of all, the recent acorn barnacles, will run to hundreds of pages. Each will take a separate volume: four volumes in all. He has finished one of the four; and the species theory is still locked in the drawer, incomplete and unpublished. He had planned to return to it by now, hadn't he? A year on the barnacles, he had promised himself. At most. Yet October would mark the end of the *fifth* year he had spent on them. They have not finished with him yet. And there is no going back.

A soft thud in the silence of the room startles him; the tiny valve in his hand falls on to the desk. First he checks the valve is safe, then turns towards the sound. He had forgotten Annie, who is now asleep on the chaise longue in his study, an honour allowed to the children only when sick. Her book, William Howitt's *The Book of the Seasons; or the Calendar of Nature*, has fallen to the floor. Darwin picks it up – she has been reading Howitt's lyrical description of January, he notes: 'Seeds are secure in the earth, or in the care of man; herbaceous plants have died down to the root, which, secure in their underground retreat, are preparing their fresh shoots, leaves, and flowers, in secret, to burst forth at spring with renewed splendour.'[2] Darwin has borrowed it from the London library for his sick daughter. It will help her to look forward to the spring, he told Emma. He closes it and places it back on the table next to her, pushing the pill bottles and thermometers to one side.

Her mouth is slightly open and her breathing heavy. The hand that was clasping the book has fallen limply over the side of the couch. Darwin steps softly. He reaches for his black Sand Walk cloak that hangs on the hook in the study and tucks it around her, closing her arm back across her body, and lifts the heavy coil of her plaited hair. Her skin is almost translucent when she sleeps, he notes, particularly against the midnight black of his cloak. The daguerreotype is a good likeness, but she has changed since it was

taken two years ago: her face is longer and narrower. With his fore-
finger he traces the familiar blue vein that branches across her
temple – Skimmed-milk White. He can see the pulse beat steadily in
her neck, now taut and stretched. There is a flush across her cheeks
– Aurora Red. She twitches, dreaming, and groans a little. Pain or
dreams? he wonders. Instinctively, he feels her forehead. It is cool.
Dr Gully's methods are beginning to work, perhaps. He stokes up
the coals in the fireplace to a brighter glow and sits to watch her
sleeping, lost in barnacle thoughts.

For a week or so Annie has joined her father in a specially adapted
version of the Water Cure that arrived from Dr Gully in the post;
and for a week or so Darwin has kept a record of her progress
alongside his own in his health diary. However, he finds defining
the subtleties of Annie's health almost as difficult as diagnosing a
species from a valve alone, or Werner's task in defining and describing colours. Words are so inadequate. So, in order to be as clear as
possible, he has made his own code, defining degrees of wellness,
better or worse than the day before, by reading the signs of Annie's
body: her pulse, temperature, bowel movements, sleep patterns and
appetite. As with Werner's colour chart or the adjectives he used to
describe the minute differences between one barnacle valve and
another seemingly identical one, the degrees of wellness could thus
be measured in a kind of chart:

well very	well almost very	well	well not quite	good	pretty good	poorly a little	poorly

This record would help Gully determine the nature of her mysterious ailment and find a cure; it was a history – a case history. It would
also chart Annie's reaction to the water treatment, so on the left of
the sheet, next to the recorded date, Darwin wrote the chosen treatment for the day – dripping sheet and spinal wash administered
every day; a wet wrap three or four times a week; the lamp treatment
once a week – and in the next column her state of health in the hours
following: 'well' or 'pretty good', for instance. Brodie and Bessie's

morning work had been much increased by this new routine, but they did not complain, though Emma had noticed that they had become less patient with Etty. Annie was not a complainer, but she cried out none the less when the cold sheets touched her feverish skin. She was more patient than he had been with the cure; and she enjoyed the warmth of the study when it was all over, too, she said.

The family health diary, Darwin reflected, was a record of oscillations – sometimes wild oscillations between the 'poorly' and the 'well' or 'well almost very'. It was not a narrative of progress – no perceptible hour hand ticking steadily towards the 'well very', or even an imperceptible minute hand, but rather more like the swings of a pendulum wildly out of kilter. It was a Foucault's pendulum, swinging erratically, like the instrument the scientific journals described on display in Paris, heavily swinging from its metal chain, proving in its slow swings the curve of the Earth's surface. Even Foucault's pendulum had some degree of predictability – like the weather; but the seasons of Darwin's own body seemed to move to no gravitational pull or position of the sun, and now Annie's little pendulum was swinging fitfully alongside his own. They shared inherited characteristics, as he had feared. All he could do was record movements and try his best to see patterns of correlation between the things they ate, or drank, or did, and the unfolding patterns of wellness. Her spasms of pain were the worst to bear – stabbing, jagged pains in her stomach that made her cry out and writhe. No one could take those pains from her; not even her father.

It had been six years since he had watched or described any of his children this closely. When the elder three had been babies, he had kept a journal exclusively to take notes on their development as they hiccupped, sneezed, cried, learned to smile, pulled themselves to a standing position or tried to work out the puzzle of their own faces reflected back in a mirror. He had been particularly interested in how they cried, he remembered – when do real tears appear? And how do the muscles of a baby's face work in crying? He hadn't written so much about baby Annie as he had about baby William – though Annie had come to be his favourite since those early writings. William had been the firstborn, after all – the first unfolding of a miracle. Babies Annie and Etty were simply points of comparison in the journal; William was the chief case study – the animalcule, as he had called his son.

Though he hadn't written much about Annie, her smile had been recorded as one of her chief characteristics from the start – that broad smile imprinted on her face as soon as her facial muscles could conjure it up: 'Annie at 2 months & four days had a very broad sweet smile & a little noise of pleasure very like a laugh,' he had written. Now, almost ten years later, he was listening again for those little cries that were yet more signs of her level of pain, writing on 24 January: 'two little cries', 27 January: 'late evening tired and cry' and next day: 'early morning cry'.[3]

Darwin had stopped writing about the children back in September 1844 – the year he had put the species theory away. Both accounts waited for the return of his attention. Both were in limbo – suspended, usurped by barnacle-watching and barnacle-writing. Now Annie was back in his writings, not as a healthy specimen of childhood development but as an ailing one. The overall pattern of her health had, he feared, shown gradual decline since the previous summer. She had not been herself, Emma reminded him anxiously again and again, since that bout of illness that had begun in the storms of the previous summer. Annie was not herself. What did being Annie mean, though? Might not these changes, these mysterious aches, this lack of concentration and fitful irritability be part of another meta-morphosis, the emergence of a woman's body from that of a child? Perhaps the grown Annie would not be so even-tempered. Perhaps the adult Annie would bear no relation to her free-swimming self.

The women of Down House – Emma, Brodie, Bessie, Miss Thorley – prayed for Annie. Aunt Sarah in the village prayed for her. Fanny Wedgwood, Emma's sister-in-law, prayed for her. Charles couldn't. He did not feel at ease about consigning Annie to the will of God. It seemed such an arbitrary will – taking, snatching where it could. He could see how nature might be so, but to try to fathom out and live with a sentient being behind and responsible for all such tragedies – that was a task for the women.[4] It was beyond him. Emma seemed to gain peace from believing she would see all her lost ones again – her baby, her sister, her parents – in heaven; but he, Charles, could not share this confidence with her. At least, not since his father died. After all, how could he be faithful to a religion that required him to believe that his father, a declared non-believer, had been consigned to hellfire? Perhaps Annie, whose belief could not, at the age of nine, be a robust one, teetered on the brink of hellfire, too, for all her goodness.

Since his father's death, in 1848, Darwin had been drifting with these problems, not quite struggling. Reassuringly, many of the memoirs he read by eminent intellectuals seemed to tell a similar story of spiritual drift, as if these books that his brother Erasmus had sent him or recommended had found him out, engaged him in silent dialogue, confirming or challenging his own ideas and anxieties. Erasmus was no believer, and seemed to embrace these radical London ideas with less conflict than Charles did; but then Erasmus had no children's souls to care for, and there were few churchgoers in his immediate circle.

Darwin continued to order each of the new books by the free-thinker Harriet Martineau, who had been such a close friend of his and Erasmus's back in the London years, before he had met and married Emma. Harriet continued to outrage some people in her outspoken commitment to telling the truth, however it looked to her. At the end of her book, *Eastern Life: Present and Past*, for instance, which Charles had read in 1849,[5] she had concluded passionately:

when all thinkers say freely what is to them true, we shall know more of abstract and absolute truth than we have ever known it yet. It is no concern of the thoughtful traveller whether what he says is familiar or strange, agreeable or unacceptable, to the prejudiced or to the wise. His only concern is to keep his fidelity to truth and man: to say simply and, if he can, fearlessly, what he has learned and concluded. If he be mistaken, his errors will be all the less pernicious for being laid open to correction. If he be right, there will be so much accession, be it little or much, to the wisdom of mankind.[6]

Many did not want to hear about her truth. In her most recent book, *Letters on the Laws of Man's Nature and Development*, which she had co-written with the phrenologist and psychologist Henry Atkinson as a radical 'correspondence between two friends',[7] she had rejected God and Christianity – in print. Christianity was no more than superstition, the letters claimed, and superstitious people refused to see the blinding truth being revealed by science. Christianity was a 'notion of the cave': 'Science is gradually leading us through these notions of the cave into open daylight, by showing the undeviating laws of nature: and thus men are gradually drawn out of the Church, into the lecture-room.'[8] And there were many others who were afraid to wake up: 'There are many who have a half knowledge that their religion is but a waking dream yet beg you will not disturb them.' But, Atkinson preached, till we 'recognise this science, we live in a barbarous and dark age, and have no health in us'.

The women around Darwin might largely have been believers – some even devout; but there were plenty of women who were growing away from religion, like himself, however reluctantly. Men too – such as Francis Newman, Cardinal Newman's younger brother, whose book, *The Soul*, which Darwin read in 1849, expressed radical new ideas about the afterlife and the prospects of Christianity. There was also Harriet Martineau's publisher, John Chapman, setting up the *Westminster Review* in London, dedicated to ideas of progress and the light of science. They were brave people, outspoken people, but it was not Darwin's way. He would keep his doubts from Emma. Such thoughts would trouble her, especially now, when she needed her faith and her God to steer her through this most troubling of times.

In early March, Annie rallied for her birthday and even rode on William's pony. From the house Emma watched her playing in the hedge with William, and it gladdened her heart. But then, two weeks later, on 13 March, Darwin's notes on Annie's health recorded a downward turn: 'Poorly with cough and influenza,' he wrote, followed by a string of dittoes for eight days. By 21 March he could no longer take the strain of watching and recording Annie alone. This was the final straw for Emma and Charles. Emma, heavily pregnant, was worn out with worry; they both were. His stalked barnacles would have to wait. He wrote to Dr Gully and arranged rooms for a month for Annie, Etty and their nurse Brodie in Montreal House in Malvern. Miss Thorley would join them if possible. Annie now needed Dr Gully's direct help – he was good with children and had several of his own. The journey with coughing Annie took two days, for they stayed overnight with Uncle Erasmus in Park Street, near Hyde Park. The following day Darwin settled the nurse and two children into the Malvern lodgings, consulted with Dr Gully and left for London two days later. He arrived in London just in time to be registered at Erasmus's house for the National Census; and in London he had something of a holiday from his barnacles and from the worry of Annie, visiting friends and going to the library.

For two weeks Darwin left the care and recording of Annie to Dr Gully and to Miss Thorley, who had joined the children and their nurse at the beginning of April. He returned to Down House and began to assemble his notes on the recent stalked barnacles for the second of the four volumes. His immediate task was to try to explain the mystifying and curious complemental males he had found on the *Ibla* and the *Scalpellum. Mere bags of spermatozoa; minute, worm-like, singular; perfect, extraordinary, wonderful.* It was difficult to speculate about their origin without expressing simultaneously the wonder and the transience of it all, for these little males lived only to fulfil their function and then dropped off their female host to be replaced by others. Again he was reminded about the usefulness of his species theory, for he would never have been led to investigate the *Ibla* and thus discover the complemental males if he hadn't already had an idea that separate sexes had evolved from hermaphrodite forms. He was plotting a bloodline in these books, starting from the ancient hermaphrodite *Pollicipes*, through the *Ibla* and *Scalpellum* to the recent stalked barnacles: diversification and variation; branching and splitting.

Meanwhile, as spring continued to unfold into April, Annie took the water treatment and seemed to stabilize, Miss Thorley reported in her letters. She went on donkey rides with them, though she still tired easily. Etty entertained her sister with collections of tiny ladybirds kept in matchboxes, which the sisters fed on milk; but suddenly and without warning on 15 April Dr Gully sent an urgent and alarming note to Darwin, asking him to return to Malvern immediately:

Annie had a high fever and was dangerously ill. Darwin dropped everything and left immediately, leaving instructions for Emma to make sure that she had chloroform in the house in the event of the baby arriving early. Two days later he arrived in Malvern, hoping to reach Annie and at least speak with her before it was too late. Etty, then a very small child, remembered her father's arrival and 'his coming in, and after Miss Thorley saying something, his flinging himself down on the sofa on his face, and Miss Thorley sending me out of the room in a frightened way'.[9] Miss Thorley, who was already at breaking point, told him Gully's prognosis: it would all be over by the morning.[10] He cried a good deal that day and over the next seven long days, for Annie did not die that night. Gully, who had stayed the night in the lodging house to be on hand should Darwin need him, conceded at dawn that if she was no better she was at least no worse. 'It is', Darwin wrote to Emma at noon the following day, 'now from hour to hour a struggle between life and death. God only knows the issue.'[11]

He resumed the narrative of Annie's sickness with no idea of its ending, writing to Emma, marooned in Down and too heavily pregnant to risk the journey, immediately on his arrival at Malvern, struggling to be absolutely truthful even if that truth might break a mother's heart: 'She looks very ill: her face lighted up & she certainly knew me. – Thank God she does not suffer at all – half dozes all day long . . . My own dearest support yourself – on no account for the sake of [ou]r other children; I implore you, do not think of coming here . . .' Then he added in a postscript, presumably added after he had seen Dr Gully: 'I am assured there is great hope. – Yesterday she was a little better, & today again a little better.'[12] He was desperate to see improvement even if the day-to-day changes were imperceptible.

Emma, renowned for her tenderness as a nurse, heartbreakingly separated from her eldest daughter, sent another pair of eyes to watch with Darwin's own – those of Fanny, who had an 'eye for illness', and four children of her own. Emma pleaded with Charles: 'You must let her experienced eye do some of the watching.'[13] Fanny, emotionally strong and organized, arrived in Malvern from

London with her lady's maid, who took Etty back to London with her to be distracted by her cousins, who had been sent to stay with their Aunt Caroline at Leith Hill Place. Fanny took lodgings next door and settled herself in. Her task was to watch over them all. She wrote to Emma about Charles: 'I do not try to prevent him doing a good deal about poor Annie. It seems as if it was some relief to be doing something, though occasionally it may be too much.'[14] Fanny also took particular care of Brodie, who was ill with worry, crying constantly and unable to sleep. Miss Thorley, she noted, had no less feeling but more self-restraint.

Through these seven sickbed days and nights, Charles, Brodie, Fanny and Miss Thorley watched over every flicker of Annie's recumbent body, sponging her limbs with vinegar, spooning broth, wine, orange juice or water into her mouth, rubbing her feet, trying to change the bed sheet without hurting her frail, thin body, and, as things got worse, holding her still while the local doctor applied a catheter to 'draw' her urine. They watched each other, too, looking for signs of more or less hope in each other, or signs of exhaustion. Together in this room filled with the smells of chloride of lime and vinegar, and full of the paraphernalia of the Water Cure, they listened to Annie's ramblings. The fever had made her delirious and she mumbled and sang for many hours at a time, then suddenly came to and recognized one of them or thanked them politely for a spoonful of orange juice or gruel. Everything had to be observed and recorded.

Although Annie's characteristically fine manners remained constant throughout the pain and delirium, Darwin was shocked more than anything by how little of the Annie he knew remained. It was a little like watching at the bedside of a stranger, until those longed-for moments of recognition. He wrote to Emma:

This morning she is a shade too hot; but the Doctor . . . thinks her going on very well. You must not suppose her out of great danger. She keeps the same; just this minute she opened her mouth quite distinctly for gruel, and said 'that is enough'. You would not in the least recognise her with her poor, hard, pinched features; I could only bear to look at her by forgetting our former dear Annie. There is nothing in common between the two . . . Poor Annie has just said 'Papa' quite distinctly.[15]

Charles cried a good deal over the next week. He cried when he read Emma's tender notes and frantic enquiries. He cried when she enclosed a pressed flower she had picked from Annie's garden. He cried after Dr Gully's visits and when Annie recognized him. Gully had explained that, with gastric fevers such as this, nothing could be determined for certain until the end of a fortnight from the onset of the illness. So they were playing a waiting game. Annie would either survive into the last week of April, or not. Each further day gave them more hope.

For seven long days and nights this routine continued. The dawn light filtered in through the curtains, gradually filling the room; shadows moved slowly; then, later, lamps would be lit, meals taken in an adjoining room, letters received and written, reports made. Annie's temperature rose and fell; her pulse strengthened, then weakened. Visits from Dr Gully would raise or dash their hopes. Each time Charles struggled to summarize Annie's sickness since the doctor's last visit: her pulse, whether or not she had vomited, and the degree of delirium she had shown during the night. Each time he struggled to be as accurate as he could be, resisting the impulse to exaggerate her pain or be overly optimistic. Gully would listen and then pronounce his sombre judgement: 'Not essentially worse,' he would say, or 'She is turning the corner,' or 'No progress.' On one occasion, Charles wrote agonizingly to Emma, he had recorded that Annie had been able to urinate without the catheter and that her bowels had moved at the same time. He had been sure that this was a certain sign of improvement and he could not wait for the doctor's arrival. He had felt 'foolish with delight'; but when the Doctor arrived, he saw these signs quite differently, as a reason for alarm, not delight. 'These alternations of no hope and hope sicken the soul,' he wrote.

In Down, Emma relied totally on the reports she received by post twice a day. She strained to listen for the sound of the postman's step on the gravel path above the other domestic noises. In the last stages of pregnancy, she struggled to keep the house running smoothly and the children calm, without the usual support of Miss Thorley, Brodie and Charles himself. Emma's seventy-year-old

Aunt Fanny came to Down House to be with her and together they tried to interpret Annie's progress at a distance, from the details of Charles's letters alone. Emma trusted Charles to be absolutely truthful in his accounts: 'I tell you everything just as it is, my dearest Emma,' he wrote, and Emma responded with gratitude: 'Your account of every hour is most precious.' She dated and ordered each letter so that she could read them repeatedly over and over in sequence in order to try to discern where they were leading: 'Your two letters of Monday are certainly better. Poor sweet little thing! I felt more wretched today than any day, but now I do think looking at the accounts of the last four days that there has been progressive improve-ment from that time.' She summarized and condensed Charles's reports of Annie's illness into her own diary accounts, writing for Easter Sunday, 20 April: 'sick 3 or 4 times & took brandy once'; 21 April: 'much better'; 22 Tuesday: 'Diarrhea came on'. That same day she received a letter from Fanny to warn her that Annie was in imminent danger and that she was to expect the worst.

Wednesday 23 April was stormy. The rain swept off the Malvern Hills, lashing at the windows of Montreal House. Darwin remem-bered how each of Annie's turns for the worse seemed to have been accompanied by storms. When she had lain in bed with Emma with influenza, the rain had fallen heavily for several days. When she had first become sick, the previous summer when the Wedgwood cousins were visiting, there had been a great storm with thunder and lightning. And here it was again.[16] Darwin might almost have felt superstitious. Brodie almost certainly was, coming as she did from the coast of north-east Scotland, where storms would often signal fatalities in the village. Some of the fisherwomen would cover the mirrors in their houses during a thunderstorm and leave the windows and doors open, so that if the thunder got into the house, it could get out without having to damage anything.

By noon, the thunder was breaking in Malvern and it was difficult to listen for Annie's last faint breaths above its roars. Charles sat down to write the worst letter he would ever have to write:

My dear dearest Emma, I pray God Fanny's note may have prepared you. She went to her final sleep most tranquilly, most sweetly at twelve o'clock today. Our poor dear dear child has had a very short life but I trust happy, and God only knows what miseries might have been in store for her. She expired without a sigh. How desolate it makes one to think of her frank cordial manners. I am so thankful for the daguerrotype. God bless her. We must be more and more to each other, my dear wife.'[17]

Emma didn't receive this letter until the following day. On the Wednesday on which Annie died she waited for a letter, but John Griffith the postman had nothing to give her. Knowing the family's grief, he was apologetic, and checked the bag a second time, but there was nothing. This meant the worst. She wrote the following day to Charles with as much stoicism as she could muster: 'Till four o'clock I sometimes had a thought of hope, but when I went to bed, I felt as if it had all happened long ago. When the blow comes, it wipes out all that preceded it and I don't think it makes it any worse to bear. . . . My feeling of longing after our lost treasure makes me feel painfully indifferent to the other children, but I shall get right in my feelings before long.'[18]

To Fanny she wrote: 'I do feel very grateful to God that our dear darling was apparently spared all suffering, and I hope I shall be able to attain some feeling of submission to the will of Heaven.' She was struggling with her God. Fanny, knowing that Emma and Charles needed to comfort each other, insisted that Charles return to Down as soon as he could, and he arrived by nightfall on Thursday. Fanny stayed in Malvern to oversee the laying out of Annie's body and to arrange the funeral with her husband, Hensleigh. She wrote to Emma and Charles on 25 April: 'Hensleigh and I are just returned from that sad & last work of laying your dear child in her earthly resting place.'[19] Later, at Darwin's request, she sent him a small map marking the position of Annie's grave in the graveyard of the Priory Church, under a cedar of Lebanon. The tombstone read simply:

ANNE ELIZABETH

DARWIN

BORN MARCH 2, 1841

DIED APRIL 23, 1851

A DEAR AND GOOD CHILD.

But these were not to be Darwin's last words on Annie. A week after her death, when he had begun to be able to control his tears, when he had answered several letters of condolence to the men of the family and announced her death in *The Times*, he took several pieces of mourning paper in order to write a final obituary of his daughter. It was an attempt to recover the Annie who had slipped away, to summon a clutch of memories and to separate off the living, joyous Annie from the dying and dead one with the pinched and sad features. How to put Annie into words? What was she now and where was she now? Who had Annie been to herself, to her mother, to her father? What had been the significance of her short life? And the meaning of it? Now, with the problem of conjuring up Annie before him, memorializing her, preserving her spirit and the memories she had left behind, he was strangely and surprisingly uninhibited. The words came easily, fluently. This was a wholly different experience from his usual daily struggle to describe the barnacles: the scratching out of abandoned and inadequate adjectives, the torn-up paper. Here, too, he needed to be precise, detailed; he needed to be able to distinguish Annie from her siblings, define her unique combination of characteristics like those of the barnacles; but the precise, passionless language of barnacle-description would not do now:

Genus: Lepas

Description: Capitulum flattened, subtriangular, composed of five approximate valves. The valves are either moderately thick and translucent, or very thin and transparent; and hence, though themselves colourless, they are often coloured by the underlying corium. These surfaces are either smooth and polished, or striated, or furrowed, and sometimes pectinated.[20]

He drew on other parts of himself to describe Annie, reaching for adjectives he had not used for years. They came with little effort, with his heart so full. He crossed out very little and, as he wrote, Annie rose up clearly and strongly before his mind's eye, like her ghost with him in the room, lying asleep on the chaise longue. He could almost hear her heavy breathing still. He could see her face above all – the translucent skin, the lengthening facial bones, the branching vein across her temple; and he could sense her spirit, the way she made other people feel when she entered the room, the little dances she would do to tease him on the Sand Walk. He could look at her still in his mind's eye, just as he remembered she had looked at him with that extraordinary directness of a child's gaze. He changed almost nothing but stopped to correct one sentence that began 'From whatever point I look at her' to 'From whatever point I look *back* at her'. He struggled to place her in the past tense, when she was so vividly there before his eyes: her sparkling eyes glinting round the door of the study as she smuggled him in some snuff, which Gully had forbidden.

21 Darwin's Memorial of Annie

In this beautiful and lyrical twelve-page account he was concerned to distinguish Annie from her brothers and sisters and to define her 'chief characteristics', the touchstone of her personality. This was easy, he found: she was joyful, loving, neat and sensitive; and he remembered her touch – her hunger for the warmth of her mother's face and longing to stroke the bare skin of Emma's arms. She loved to be kissed. She loved to smooth her father's hair and his clothes. Now that she was gone, he felt her death as an emptiness where she might have been – an ache. He could conjure her up in his mind's eye, see her working at his desk or in the schoolroom, comparing the shells or feathers she had collected with his colour chart – even smell her sometimes; but he could not feel her. He would reach for her face or the shine of her hair and find only empty space. 'Her figure and appearance were clearly influenced by her character: her eyes sparkled brightly; she often smiled; her step was elastic and firm; she held herself upright, and often threw her head a little backwards, as if she defied the world in her joyousness. Her hair was a nice brown and long; her complexion slightly brown; eyes dark grey; her teeth large and white.'[21] 'But looking back,' he concluded, 'always the spirit of joyousness rises before me as her emblem and characteristic: she seemed formed to live a life of happiness.' Annie was made for joy – reason enough and meaning enough for a short life.

Horace was born three weeks after Annie's death. Perhaps Emma saw Annie there in the flickering fluidity of her newborn son's face – a gesture, an expression that for a moment brought Annie back. Nothing is ever quite lost. She struggled to come to terms with her God, who had given Annie life as well as taken it away. When it seemed most unbearably unjust, she had to explain her God to Etty, answering Etty's persistent questions about where her sister was with as much confidence and affection as she could. Annie was now with the angels, she told her. Etty, seven years old, struggled to map out the place that Annie had gone to and became confused. If all the angels were men, like Gabriel, then where would Annie have gone? Were there separate rooms for the girl angels? she asked. She wanted to be good, afraid of the consequences if she was not, for the maids had always compared her with the good Annie, the angel sister. If she wasn't good, she might be left behind when her mother and all the good children went to heaven together. She pleaded with her mother to help her to be good: 'Mamma, I used to be a very naughty girl when Annie was alive. Do you think God will forgive me? I used to be very unkind to Annie.'[22] Emma assured her that they would all be reunited in heaven; but Etty knew that something did not make sense: why had Jesus taken the good one?

Charles kept silent. Perhaps he longed to relieve Etty of her tormented resolutions to be good, seeing in them his own struggle as a child after his mother's early death, his own struggles as a man with the death of his unbelieving father. Annie had indeed been good; but goodness had not helped her to survive. Why had she died? Because she had inherited poor health. She had become too weak to survive. Perhaps he longed to tell Etty that she did not need to worry about watching herself, recording herself, being good enough for God. Perhaps he wanted to say what he was beginning to feel himself: that these were all 'notions of the cave' and that after death there was nothing – no God waiting to scour Annie's record book to decide whether she would be consigned to heaven or hell. She had only to be strong for life; and she hadn't been strong enough.

> One thing is certain, that Life flies;
> One thing is certain, and the Rest is lies;
> The Flower that once has blown for ever dies.[23]

9

Corked and Bladdered Up

*

Since I have undertaken to manhandle this Leviathan, it behoves me to approve myself omnisciently exhaustive in the enterprise; not overlooking the minutest seminal germs of his blood, and spinning him out to the uttermost coil of his bowels.

Herman Melville, *Moby Dick*

By the end of May 1851 the Down household had accommodated Annie's death, gathered its sorrow back into the patterns of its days. The newborn baby, Horace, whose arrival Emma had longed for, absorbed his mother's days and emotions. William, now eleven, boarded at the local prep school at Mitcham in Surrey, one of six boys tutored by the Revd Wharton. So Etty (6) now found herself the eldest of this household of small children: George (5), Elizabeth (3), Francis (2), Leonard (1) and the newborn Horace. There was a new nurse, too, for Brodie, no longer able to bear Down House in the wake of Annie's death, had asked to retire and had returned to her family on the north coast of Scotland. The life of Down closed around them sympathetically. In the woods where the Sand Walk curled, forget-me-nots glimmered, staining the undergrowth a delicate pale blue like the blackbirds' eggs in the nests of the hedge. The migratory birds had returned and the swing had been hung up again in the yew trees in Down House garden.

Charles turned gratefully back to his routines and to the busy life of Down, seeking to fill the gap that Annie had left behind, keeping busy. As he walked around the village, the men from the local Friendly Club who played cricket in his meadow, the carpenters and blacksmiths who had worked on his house and the village shop-keepers passed on their condolences. At home the house was full of new life – the two little boys, Francis and Lenny, for whom Annie was only a passing shadow, no more important than the nursemaids who chased them, played in the garden as before, running after butterflies and catching ladybirds. Charles lay out under the big trees, feeling the warmth of the sun on his face, while the little boys climbed over him, pretending that he was a mountain bear, running their hands though the thick hairs on his chest and arms. He heard their shouts and laughs when he sat in his study. The frame of the study window was filled with startling green – the fresh leaves of lime trees. There were letters to answer and a book to finish.

The letters Darwin received in May were written to a man of authority; he was older and wiser. Young naturalists had been writing to him for some time now, on the strength of either his *Beagle* book or his coral papers, or because one of the London professors such as Owen or Forbes had passed on his address. They wrote simply to tell him that they had found a rare fossil, or to ask him to read a draft paper, or for a job reference. There were many appeals to his authority and knowledge that summer.

A young geologist, J. S. Disnurr, wrote to him in early May to ask him for advice on some fossil footprints he had found near Port Philip in Australia. Darwin advised him to go to see Richard Owen at the Royal College of Surgeons in Lincoln's Inn Fields – he would verify the find. Disnurr did as Darwin suggested, reporting despondently a few weeks later that Owen had discovered that they were not fossil footprints at all but simply hollows left in the rock by nodules of iron. Darwin felt for his embarrassment, remembering his own blunder in relation to the roads of Glen Roy, and he tried to lift the spirits of the young explorer before he returned to Australia on his ship: 'Everyone makes plenty of blunders at first & I well know that I have done so – & so long that they are not printed & published, it signifies nothing.'[1] Send me your specimens, he suggested, and I will find 'qualified people to appreciate them & describe them'; but his advice above everything was to be careful – don't publish anything until you are absolutely sure of your facts and have had them verified by 'qualified people'. It was a cut-throat world out there.

There were some blunders, however, that Darwin found it much harder to forgive, from the new generation of headstrong and ambitious naturalists and explorers, the new blood that competed for publication and recognition, some of them with private incomes like himself, others pursuing their sponges or fossils or seasquirts while maintaining a job, like Fitch the Norwich chemist or even Bowerbank the distillery owner; and a rare few competed for the handful of low-paid jobs in universities. So many, Darwin realized that summer, were working on sea creatures; marine zoology dominated the proceedings of the Zoology section of the British Association for the Advancement of Science meetings. It was not surprising, perhaps – there seemed to be a consensus now amongst zoologists that the answers to the origins of life lay in the riddles of rock pools and seabeds.

Among the piles of still-unanswered letters of condolence from family and friends on Darwin's desk in May was one from Edward Forbes, Professor of Botany at King's College, London. Forbes was himself under a stone; he had been occupied since 1848 with writing up his life's work: a classification of British molluscs in collaboration with Albany Hancock. He was still working on the third of the four volumes. Forbes sent Darwin his sympathy, but also, perhaps in an effort to distract his friend from his grief, asked him to assess a barnacle paper submitted for publication by a young dentist from Swansea called Bate. Forbes was effectively asking Darwin to act as a referee. There were few people in England publishing exclusively on barnacles except Albany Hancock, who was still worrying about burrowing barnacles, and John Gray, the keeper at the British Museum, who had politely stepped aside from his barnacle research when Darwin began his work. Here was a new voice – a man he had never heard of. Darwin began to read this paper with both trepidation and excitement.

Darwin was shocked by the messy piece of barnacle scholarship that lay before him. The young Bate was clearly enthusiastic enough. He was also a fine draftsman who had done some original research; but he simply didn't know the field and, worse still, he claimed barnacle discoveries that others had made before him. Replying to Forbes's letter, Darwin could barely contain his outrage within the conventions of politeness and his own natural generosity: 'I am sorry to say that Mr Bate is not at all aware (as he suggests him-

self) how much has been published on the Cirripedia,'[2] he began
with restraint. In the remainder of the long letter to Forbes, Darwin
outlined his objections and the catalogue of Bate's omissions and
blunders, but in doing so he began to realize the scale of his own
knowledge. The barnacles had been in chaos, he remembered,
when he had begun five years before – complete chaos; but since
then others had made significant discoveries and it was now possible to say that advances had been made. The field, though small,
had changed rapidly since he had entered it in 1846 and he was
bringing it all together. It gave him an opportunity to confirm to
himself that he had reached a place at the top of the mountain of
barnacle knowledge, able to survey the territory before him; young
Bate was merely looking up at the calcareous foothills.

Yet the dentist was considering going to press. He clearly thought
he had discovered and illustrated the very first stage of the metamorphosis of the larvae of stalked barnacles, but the German Karl
Burmeister had discovered that back in 1834 – nearly seventeen years
before. Then there was Henry Goodsir's work on the metamorphosis
of sessile barnacle larvae published in 1843. This was all old hat. Bate
wrote as if mature cirripedes were blind, but the American Professor
Joseph Leidy had discovered eyes in mature barnacles in 1848; he also
seemed to think he was the first to illustrate barnacle sperm, but Dr
Rudolf Albert von Koelliker of Wurzberg had drawn them – admittedly not well – in a published paper of 1843. Bate's drawings of the
sexual anatomy of barnacles were ludicrous too: he had drawn the
alimentary canal leading to the end of the barnacle penis and his
knowledge and illustration of the barnacle female parts, wrote
Darwin, 'are *very* far, as I believe from the truth'. To understand
female barnacle reproductive parts Bate would need to read the work
done in France in 1835 by Gaspar Joseph Martin-Saint-Ange and by
Professor Sven Loven in Scandinavia. He himself, over five long
years, had verified and in some cases modified all of these discoveries.
Bate was in the barnacle dark ages. Darwin suggested that Forbes tell
Bate to read the works he had listed and revise his paper in the light of
this reading, also revising his claims to barnacle discoveries. 'It would
be unfriendly not to caution him,' he told Forbes, darkly.[3]

Darwin was always exacting, but generous to younger naturalists. Within a few months of this exchange with Forbes about Bate's blunders, he and Bate were in direct communication and exchanging findings. By way of thanks and apology, Bate sent Darwin his drawings of the second leg of the larva barnacle in different stages. They were good and careful drawings, and Darwin admired them. The older man offered his empathy for Bate's disappointment in discovering that his work was not new; such things happened to all scientists, he said: 'It has occurred to me before now, to have been working hard at a subject, & then found that my results had been previously published, & very much provoked I have felt. – therefore I can appreciate & admire the very pleasant manner in which you received my unpleasant tidings.'[4]

This experience of embarrassment at a misinterpretation or mistake, or a sense of anxiety that a new discovery might be about to be published by someone else, made him continuously humble and generous, even to someone who had blundered as extensively and carelessly as Bate. Afraid that Bate might be discouraged from further work by this disappointment, Darwin sent him valuable advice on dissection techniques. After all, Bate might be on the foothills, but with help he might make some considerable headway up the lower slopes without threatening Darwin's own discoveries. He lived in Wales, had access to the sea and to excellent microscopes, and was progressing fast with his theoretical knowledge. Darwin told him how to preserve precious and minute barnacle body parts. He described how he would take a barnacle jaw, for instance, and place it in water, then place the drop of water on a glass slide and with gold size draw a circle around the drop, sealing the slide with a glass cover slip placed on top: 'Every cirripede that I dissect I preserve the jaws &c. &c. in this manner, which takes no time & often comes in very useful. This very day I have been using preparations thus made two years since, & they are perfectly clear & with some colour preserved.'[5] He also told Bate to buy a glass-ruled micrometer to slip in the eyepiece of his microscope in order to 'measure to the twenty-thousandth or less of an inch, without delaying your work half a minute'.[6]

He enjoyed being able to pass on his practical knowledge. His neighbour, the wealthy astronomer, mathematician and banker Sir John William Lubbock, who lived in the big house, High Elms, had a seventeen-year-old son who had been visiting Darwin for two years now for lessons in dissection and preservation techniques. He would be back from Eton for the summer and Darwin was keen to show him some new techniques he had recently perfected. It was a strange rite of passage this: he had been seventeen when he had started dissecting with Robert Grant in Edinburgh. They had even carried their microscopes down on to the sands of Leith, so that they could study the sponges in situ, in the rain, in the fog, at dawn, even at night. Now he was the master and John Lubbock the pupil. Just as Grant had used him and Coldstream as skilled sponge collectors, he now had a global army of barnacle collectors of his own to draw on. Also like Grant, in return he would pass on his unique knowledge and skills and introduce younger men to other specialists. Disnurr, for instance, was going to sea again in June and he wrote to ask Darwin for advice on dredging. Darwin referred him to Edward Forbes in the Jermyn Street Museum of Practical Geology: 'Prof Forbes knows more about dredging than all the other naturalists in Europe put together,' he wrote.[7] Back in 1835 he had written to John Coldstream in Leith for advice on making a dredge; he remembered the letter and the diagram Coldstream had returned. He had also visited Grant in London to be brought up to date with the latest techniques in the dissection of invertebrates. That had been twenty years ago.

Where was Grant the brilliant anatomist and transmutationist now? Still in the chair of Comparative Anatomy and Zoology at University College, but no one had heard anything of him for years – except his students. Some said he was living in a slum in London and that he was almost completely deaf. He had never written that ground-breaking book that he had always said he would. He hadn't turned the world upside down with his invertebrate discoveries. The University had seen to that, with soul-destroyingly low pay and conditions.[8] He took home £39 a year from his post. Even Richard Owen, Hunterian Professor, only took home £300 a year before his Crown pension and grants; and Owen had seen to it that Grant was discredited too, with his distasteful and outspoken Lamarckian ideas. With a word here and there he had seen to it that Grant was isolated and eventually ostracized from the Zoological Society. This cut him off both from the zoological community and from the dead animals that he needed to continue his work. Owen had patrons and contacts and power; he was not a man to cross.

A private income and a home outside London had bought Darwin some degree of independence from these complicated networks of power. When he had returned from the *Beagle* it had been his father's wealth that had granted him the time to sift through his ship-load of specimens and to publish his papers before others did. Now in London he could see a handful of brilliant young naturalists disembarking from long world voyages as he had done, with ideas and theories and cases of carefully preserved evidence. Like Dick Whittington they had come to London, imagining it paved with gold – fame for the taking; but without a private income and patronage their theories were as nothing. Thomas Huxley, the brilliant naturalist and ship's surgeon with the hawk-like eyes who had just returned from the *Rattlesnake*, was living in rooms on Regent's Park, looking around for favours, references, patrons and piecemeal jobs. He had a fiancée in Australia, they said, and wanted to bring her to England, but couldn't do so until he had a regular income. There were few prospects in science.[9]

Boxes of barnacles came and went almost weekly at Down House. Although Darwin had returned some of the larger fossil collections to their owners, men like Bowerbank and Fitch, more were coming in. So the study was like a railway station filled with boxes either being shipped up to leave or having just come in, waiting to be unpacked. Everything had to be labelled and ordered or Darwin would lose rare specimens. Sometimes he requested particular barnacles from particular collections; sometimes people just sent them. Of course they didn't know which barnacles he was working on at any one time; they just knew he was 'doing' barnacles and offered to send him their collections, or just sent them. So they arrived randomly – he'd ask for a particular *Ibla* specimen in order to finish a section of his manuscript or in order to settle a point of fact, and the *Ibla* would come tucked into a box with a *Scalpellum* or a burrowing barnacle. So questions would open up again that he simply couldn't think about until he had finished with this immediate and particular set of problems. So he had to make himself label them, jot down a note or two for future reference and to jog his memory, and put them to one side.

At least he was beginning to feel secure in the knowledge that there were fewer and fewer barnacles arriving that he *hadn't already seen*. Robert Ball, the Director of the Museum at Trinity College, Dublin, wrote to him from Ireland in late May with the offer of a whole series of specimens that four years before would have been a veritable feast, but now he knew he had seen and mapped almost all of them. So he didn't need to see them, and he told Ball with a note of polite desperation: please don't send me your notes or your barnacles or the cast of the turtle shell: 'I do not think a cast wd be worth sending . . . many thanks for your offers . . . but . . . I do not think it likely there would be anything new.'[10] Ball sent the barnacles anyway, and the cast of the turtle shell, which he said had been 'mined' by barnacles. Darwin was amazed: these barnacles had burrowed so far into the turtle shell that they had penetrated not only bone but also through to the soft body of the turtle. Had they burrowing powers after all? Or had the bone and skin grown up around them? Why had they developed these powers? He wrote to Ball: surely 'a barnacle could not have lived in such a chamber as you describe, not openly connected with the sea-water' . . . 'I can only repeat that I am quite confounded on the subject.'

These were questions he had begun to ask on the Chilean beach sixteen years before – he had only been twenty-six when he had picked up Mr Arthrobalanus and wondered about why and how this species had evolved burrowing powers as a survival strategy. Now he was forty-two. But whilst these puzzles pressed him in May 1851, his schedule would not allow him to begin work on the sessile barnacles until 1852. Mr Arthrobalanus and his tribe would have to wait longer in the wings.

The barnacles he received by parcel carrier in the summer of 1851 came from every corner of the globe, including Angola, South America, California and South Africa,[11] but there were others he still wished either to see for the first time or to see again. When

Disnurr, who was about to travel back to Australia, offered to do some barnacle-collecting for him, Darwin leapt at the opportunity to commission the naturalist:

I am really obliged & flattered by the wish you so kindly express to send me something interesting – There is a little pedunculated cirripede I believe common on the whole South Australia –[12] Which I should be very glad to have several specimens of all sizes still attached sent home, having been placed immediately in strong spirits – well corked & bladdered up – They consist of 4 little bluish valves mounted on a flexible peduncle crossed with yellowish spines – There is a *most extraordinary* anatomical peculiarity which I want to dissect. There is also another cirripede attached to corallines of So Australia in *deep* water of which I enclose a rude tracing which I should much like to have several of for same purpose in spirits – I am at present hard at work on a Monograph on the Anatomy & Classification of all the Cirripedes (Lepas Balanus &c) in the whole world.[13]

Disnurr was footloose and fancy-free and determined to make dis-coveries, just as Darwin had been in his twenties; but now Darwin was well and truly grounded in Down – by the weight and complexi-ty of his work, by his family commitments and by his health. These naturalists – these recruits – must work for him; he must be content to be the still centre, to whom others returned with their zoological plunder, to whom others wrote for knowledge and understanding: barnacle hunters.

However, even a grounded man might travel – in his mind's eye. He could still read about the travellers, still go with them up their mountains and across the plains. In May he read George Gordon Cumming's *A Hunter's Life in South Africa* and Herbert Edwardes' *A Year on the Punjab Frontier*. He relished these tales of adventure in new lands – always had. They coloured the way he saw things, like reading Humbolt's travel stories on the *Beagle*: 'he like another Sun illuminates everything I behold', he had written.[14] Cumming's book was a bloodbath, an account of days spent shooting great game on a massive and ferocious scale, compulsively, triumphantly, for five years. It reminded Darwin of the hunting forays he had taken inland on the *Beagle* voyage – the day he had ridden across Brazilian plains ostrich hunting with wild soldiers and eaten roasted armadillos as the sun set. Cumming described shooting elephants, ostriches, giraffes, alligators – anything that moved – with single, or sometimes multiple, shots. It had taken nearly thirty shots to bring down one elephant, he wrote. Then his guides began the ritual of skinning and disembowelling the dead animal for a feast, even climbing into 'the immense cavity of his inside . . . and handing the fat to their comrades outside until all is bare . . . the natives have a horrid practice of besmearing their bodies, from the crown of the head to the sole of the foot, with the black clotted gore; and in this anointing they assist one another, each man taking up the fill of both his hands, and spreading it over the back and shoulders of his friend.'[15] Cumming was not interested in geology, natural history, anthropology, ethnography or politics; he was only interested in the volume of his trophies. He returned to England after five years of sport with thirty tons of animal skulls and skins, some of which he exhibited at the Great Exhibition.

Herbert Edwardes, author of *A Year on the Punjab Frontier in 1848-9*, which Darwin read that summer, was quite another species of adventurer. An educated military man, he was interested in politics and British overseas policy. His aims were not to boast of hunting victories but to tell the story of the 'bloodless conquest of the wild valley of Bunnoo', to give an insight into the life and labours of an Indian political officer and to pass on his knowledge of Indian life and customs. His story would be none the less exhilarating, he promised his readers, asking them to:

be prepared to enter with me on more stirring scenes; to march with me once more towards the western frontier of the Punjab; assist me to fix the yoke on the neck of a savage people; help me to turn the assassin's knife; swim with me the midnight ford, and wake the sleeping border rebel in his lair; read with me, with indignant sorrow, the betrayed and wounded magistrate's appeal for help; sound with me the loud alarum, beat the angry drum; welcome the fierce but unfriendly warriors, that rally to the call; and weld tribes, that never met before in friendship, into a common army, with a common cause; then, confident of right, plunge into THE WAR.[16]

Such men were heroes. Darwin, too, had been a hero to many readers of the *Beagle* voyage, a man who could take others to unexplored territories, who could make them see, and extend their horizons.

Hooker, recently returned from India with his triumphant stories of kidnapping and war and weeks spent in imprisonment, all for the sake of his rhododendrons, was a hero – but still a hero without an income, a hero who had to find a job. Frances Henslow had waited for him during those long years, and now they were married and honeymooning in Paris. These biographies of exploration, and the tales of the recently disembarked naturalists such as Hooker and Huxley, brought back for Darwin the days he had spent on horseback in the South American deserts: the danger, the political intrigue, the discovery of peoples living their lives so differently from those in England or Europe, the zoological and geological differences and the sublime and awe-inspiring variety.

His own vision might have been myopic that summer, completely absorbed by the delicate and minute body parts of *Ibla* and *Scalpellum* under his microscope, but the newspapers and his library books kept his imaginative mind stimulated by tales of international progress and scientific advance. Gold had been discovered in Australia, Livingstone had reached the Zambezi and the great telegraph cable was being laid across the seabed of the Channel. London seemed to be a hive of activity. The papers described the honeypot that was luring millions into London – the great glass structure of the palace that housed the Great Exhibition, designed to show the world the power, scale and inventiveness of British industry. Trains carried tourists from every great city of Britain into London. Many were travelling for the first time, travelling in family groups to see the displays of British success and diversity in trade and the arts; but whilst Emma was enthusiastic to visit, Darwin feared that London would make him ill again. He would consider it; he would need a holiday when he finished this wretched manuscript. He'd need to give his eyes a feast, a spectacle on a grand scale, to stretch his vision to bigger sights than the microscopic structures of his creatures. A holiday might give him an incentive to finish. So throughout June he laboured away at organizing the woodcuts for the volume and in trying to reach a conclusion about the evolution of these strange complemental males in *Ibla* and *Scalpellum*.

Try as he might he could not summon the linguistic certainty he needed when he tried to reach *Ibla* and *Scalpellum* conclusions. Blunders haunted him. Perhaps it was better to hedge, to defer, to make his thinking processes visible so that others might see the problems in his logic, if there *were* problems in his logic. He wrote the title out several times: *Summary On The Nature And Relations Of The Males And Complemental Males, in Ibla and Scalpellum.* Here, his speculations all depended upon a sequence of ideas and premises, like a chain in which any of the links might be defective. He decided to show the sequence – allow the reader to follow the links in long chain-like sentences, where clauses clanked against each other like steel: 'It should be observed that the evidence in this summary is of a cumulative nature. If we think it highly, or in some degree probable that [A, B, C and D] . . . if from these several considerations, we admit that [X, Y and Z] . . . then in some degree the occurrence of parasitic males in the allied genus Scalpellum is rendered more probable.'[17]

Probable; possible; likely – it was the best he could do when there was crucial missing evidence, facts that he could guess at but which he had not actually seen with his own eyes. He continued to forge hesitant phrases outlining decent probability, appealing to common sense: 'It was hardly possible that I should be mistaken . . . the only possible way to escape from the conclusion . . . a conclusion hardly to be avoided.' He was back to dangerous speculation and speculation was what this barnacle work had been about avoiding, but it was the best he could do here in one of the most baffling corners of his work. Over unimaginable stretches of time, *Ibla* and *Scalpellum* had evolved new reproductive methods to maximize their survival prospects, moving away from hermaphroditism to separate sexes and a division of labour. Barnacle variations: millions of years in the making and infinitely more astonishing, inventive and diverse than the range of breathtaking designs and productions of British industry housed in the great Palace of Glass.

The second volume of the stalked barnacle monograph was to be published by the Ray Society, a new publishing house established only seven years before, named after John Ray, a seventeenth-century naturalist, and dedicated to publishing specialist books on British flora and fauna. This was a prestigious list, but the books would only reach other specialists who subscribed to the Society, not the general reader. By July Darwin was in weekly correspondence with the President of the Society, Edwin Lankester, finalizing details of typeface and size of type, length and number and type of illustrations. His stalked barnacle volume would be number 21 in the Ray Society list, next to number 20, a book on lichens. Now that the volume was finished, he worried that the manuscript was almost impossible to read in places, but he hoped that if the printers had tolerated the previous manuscript they could make this one out. He would just have to check the proofs very carefully.

Even if the manuscript was atrociously bad, it was the *only* manuscript, and delivering it to Edwin Lankester in London would be risky. Disasters had happened to other great books such as Carlyle's first copy of *The History of the French Revolution*, which had been accidentally and famously thrown on the fire by one of John Stuart Mill's maidservants, thinking it waste paper. Everyone Darwin knew seemed to have a story to tell about missing pages or lost pages or spoiled pages. So he would take no risks and send the manuscript to London with Parslow. Parslow could be trusted to carry the package of precious paper and put it directly into Edwin Lankester's hands. He would ask Parslow to wait while Lankester looked over the manuscript, and then he would be instructed to deliver it by hand to the printers at 22 1/2 Bartholomew Close. A good plan; but Lankester wrote to say that he would need at least a week to look over the script, and so Darwin decided that there was no option but to take it himself. He would use the opportunity to have a holiday with Emma, George and Etty in London. It would do them all good to see the Exhibition and some of the London sights.

On 24 July Darwin wrote to an old dissecting companion who lived near Hyde Park in London, George Newport, for a pair of scissors. While in London he wanted to visit the surgical instrument makers on the Strand, he explained, 'in order that I may shame Mess Weiss & Co to endeavour to make me an equally good pair (but to open with a spring & mounted with one arm long, for I have in vain endeavoured to cut in the wonderful manner I saw you do with one elbow pointing at the sky) . . . Weiss has made me two pair, but they are *very poor* articles.'[18] He would have been back for more dissection training before now, he wrote, 'but I have found the excitement of London so injurious, that I have seldom come up'.[19]

On the afternoon of Wednesday, 30 July Charles, Emma, Etty and George arrived at Darwin's brother's house at 7 Park Street, Grosvenor Square, London. The journey from the outskirts of London to Grosvenor Square had taken them much longer than usual, for the streets were full of people and carriages, families who had travelled to London from all over Great Britain and the world for the Great Exhibition. Many Americans had also travelled in order to see the eclipse that had taken place two days before on 28 July and which, the papers promised, would be particularly visible from England. As it happened, the eclipse had been a rather unremarkable affair except for those few people who had watched from remote Scottish islands, the papers said now.

The lodging houses were full to bursting point. Everywhere the Darwin children spotted posters and flyers on shops and the sides of omnibuses that advertised the Exhibition and other summer events and spectacles: Wyld's Monster Globe in Leicester Square, Cremorne Gardens, the Royal Hippodrome, Madam Tussaud's, Drury Lane, The Lyceum, and the Zoological Gardens.

Carefully packed in Darwin's hand luggage by Parslow were two pairs of scissors, his own defective pair and Newport's perfect pair, a microscope and a manuscript. Darwin had already prepared a cover letter to Edwin Lankester of the Ray Society and this was packed in the barnacle manuscript parcel with the etchings. While Emma supervised the unpacking, and Uncle Ras entertained the children, Darwin walked from Erasmus's house along Oxford Street and Regent Street towards Old Burlington Street to deliver the manuscript personally.[20] The wide streets were bustling with activity, sounds and smells and sights that Darwin found he could scarcely take in: rows of carriages drawn up on at the edge of every pavement; men and women walking, on horseback or in carriages; shopmen, trimly dressed, stepping out to meet customers or to deliver parcels into waiting carriages; bakers and confectioners with open windows filled with buns and tarts; coffee shops, trunkmakers, hosiers, fruit and vegetable stalls loaded with cabbages and cauliflowers, a fishmonger drenching his shop with cold water in the heat of the sun.

Although Darwin had promised himself that the manuscript would be delivered before the end of July, this was not a good week for Lankester to look at it, nor for Darwin to be in London, if he expected to see old friends. Lankester was due to sail to Paris at any moment as one of the jurors of the Great Exhibition invited by Louis Napoleon, the President of the French Republic, to a festival in honour of the Exhibition. Joseph Hooker and his new wife Frances would be there, too, with the other British guests, who included the members of the Royal Commission, the Lord Mayor and aldermen of the City of London. Lankester would be busy with the daily schedules of glamorous receptions, fetes, theatre and ballet performances. However, Lankester had made an agreement to look at the manuscript, Paris or not, and Darwin would hold him to it. He would return for the manuscript in eight or nine days, he said. There would be no settling down to the new barnacle volume until this one was in print.

So with the manuscript safely in Old Burlington Street, Darwin and Emma could look through the papers and decide which sights they would see with the two children. Eras, living as he did in the heart of fashionable Mayfair, had seen many of the sights already and recommended some of the panoramas and lectures at the Polytechnic on Regent Street – they would be of particular interest to his science-minded brother. Emma's brother Hensleigh and Fanny had already been to stay with the hospitable bachelor Erasmus that summer and he had some feel for what Fanny's children had enjoyed. There was even a Chinese junk moored on the Thames, he told the children, and the Chinese sailors threw wild parties with fireworks at night.[21] First of all, though, they must see what the world and his wife had travelled to London to see: the Palace of Glass in Hyde Park – the Great Exhibition. Erasmus ordered three carriages to take them all the short distance from Park Street to the Prince of Wales Gate on Kensington Road, skirting the perimeter of the great park. The carriages would keep them a little separated from the crowds, he said, and save the children's feet. From Kensington Road the children peered through the carriage windows and through the crowds to get a glimpse of the fairy palace, which looked for all the world like a great cathedral made of crystal.

22 Interior of The Crystal Palace

As Charles, Emma, Eras and the children with Parslow and the ladies' maids approached the main entrance, the arched transept towered above them, catching the colours of the sky and clouds like an enormous mirror, a hundred feet high. The effect of the exhibition was 'more than sense can scan or imagination attain', *The Times* had eulogized and for Darwin, accustomed to microscopic vision and easily moved to wonder and awe, this must have seemed a vision. His eyes could simply not take it all in at once. Stretching away from them on either side as they queued to enter, the building faced them with its 1,851 feet of glass, probably the longest building they had ever seen. Through the turnstile, past the stalls selling catalogues and the umbrella stands, they stood to absorb their first impressions, suddenly stilled along with the scores of other visitors who had entered with them. Charles or Erasmus lifted the little boy George to see it all: flowers and palm trees, the towering Hyde Park elms now under glass at the far end of the transept, the brilliant blue-and-white colour scheme and the great netting of steel that held it all together between them and the sky. Before them sparkled the famous Osler's Crystal Fountain, twenty-seven feet high, made of coloured glass and surrounded by sculptures of gleaming white marble.

Through the crowds they glimpsed the glitter of the Koh-i-noor diamond, heard the thunder and hiss of the great steam engines, tasted the savoury pies and jellies in the refreshment courts, looked up at the stuffed elephant, down at the scientific instruments, microscopes and inventions, wandered through the stained glass and carved wood of the Mediaeval Court, peered into tented courts of Indian furniture and textiles, admired artefacts made of glass, crystal and steel by British manufacturers, and heard the incessant hyperbole of the attendants about the might and inventiveness of British industry. Like most visitors, they were overwhelmed by things – by light, by invention and by progress. Eventually it was all too much. Darwin's head swam. The children became cross and quarrelsome; Emma tired. Erasmus had seen this all before with his other visitors: Exhibition fatigue; time to go home.

Another day Emma and Darwin left the children in the care of their uncle to visit other sights: the Overland Mail diorama at the Gallery of Illustration at 14 Regent Street. The advertisement in *The Athenaeum* promised: 'The Diorama of the Overland Mail to India exhibiting Southampton, the Bay of Biscay, Cintra, the Tagus, Tarifa, Gibraltar, Malta, Algiers, Alexandria, Cairo, Suez, the Red Sea, Aden, Ceylon, Madras, Calcutta, and the magnificent

Mausoleum of the Taj Mehal, the exterior by moonlight, the beautiful gateway and the gorgeous interior, lighted by crystal and golden lamps'.[22] All in one afternoon. Together he and Emma would travel to India without leaving their seats, by watching a revolving series of vast brightly lit, richly painted scenes moving across the stage in a darkened theatre, reconstructing the journey made by Thomas Waghorn, a naval officer, between London and Calcutta. They were used to mental travelling, reading as they did, listening to Emma's piano playing as they did – they could conjure up the same images. Emma had travelled at sixteen throughout Europe with her sisters – the Grand Tour – but she only knew the East through Harriet Martineau's descriptions. Now it was all so vivid and sublime in the moving diorama images, new exotic worlds opening out before them.

23 Royal Polytechnic Institution

At the Polytechnic Emma and Charles listened to lectures on the chemistry of the minerals of the Great Exhibition, including the Koh-i-noor diamond; on the total eclipse of the sun; demonstrations of Foucault's Pendulum, accompanied by explanations of how the heavily and apparently erratically swinging pendulum proved the curvature of the Earth's surface; a lecture that promised to explain how newly invented gas cookers would revolutionize British kitchens; a lecture on the history of the harp with vocal illustrations; and watched two series of dissolving views of a diver and diving bell – all of this for only one shilling: a bargain. They also took the children to the Zoological Gardens, which were an entirely different place now that they had opened up to the public, Darwin noted. It was astonishing that these animals drew crowds as large as this – there had been over 300,000 visitors since January, the porter told him.[23] They saw the family of elephants, monkeys in small and cramped cages, the young orang-utan from Borneo, the new and depressed alligator, which refused to surface, horned lizards, hummingbirds, ospreys and jaguars, and the famous hippo Obadiah, which Annie had seen the previous year with Miss Thorley.[24] In a week Emma and Darwin had not left London, but they had travelled to India, Africa, Greece, Egypt and to the bottom of the sea – all by means of panoramas and dioramas and a garden full of exotic animals. Like millions of visitors to London that year, they found in London all the empire and exotic lands they needed, carefully performed at their convenience in the comfort of Leicester Square or Piccadilly.[25]

Darwin went shopping, too; but unlike Emma's shopping list, which may have included some of the new fabrics she had seen at the Exhibition, or new books from the London Library, Darwin's list included scissors and gold size. He visited Smith and Beck, the suppliers of his microscope, on Colman Street to renew his supply of the gold size that he used for sealing his barnacle specimens between glass slides. Mr Smith was keen to show his eminent customer that asphalt might do a better job. It came in one-shilling bottles, he explained, lasted longer, was easier to apply and formed a more effective seal. Darwin bought several bottles.[26] Some time in this week, Charles walked alone to the Strand, carrying his two pairs of scissors. Weiss and Co. at number 62 The Strand were indeed shamed by the complaints of one of their most famous customers. They took the two pairs of scissors and promised their best service. They listened to Darwin's passionate and frustrated explanations about how he needed springs so that he could dissect with his elbow to the sky, and they admired the scissors he had borrowed from George Newport; but, Mr Weiss pointed out, Darwin's own scissors perhaps only needed a little adjustment and sharpening. They were, after all, almost identical to Mr Newport's. Darwin had to agree; he would not need to buy a new pair if Weiss could achieve some magic with the polishing and sharpening machines.[27] So he left his precious scissors in the Strand, just as he had left his precious manuscript in Old Burlington Street: trusting.

Darwin walked back along the Strand past the scientific and optical instrument shops and past John Chapman's publishing house at number 142, where billboards advertised second-hand books from Chapman's first catalogue. Chapman had published so many of the books Darwin had read recently: Francis Newman's books and Harriet Martineau's dangerously atheistical latest publication. Chapman seemed to be the daring centre of a radical coalition; his Friday night soirées were attracting attention across London, Erasmus said. So Harriet had found her radical non-conformist coalition, her outspoken family; and behind these doors that opened on to the Strand and above the shop, the life of this household was a daring one, Erasmus gossiped. Chapman's house was always full of visitors and lodgers, particularly this summer, and some said that his wife Susanna ran the house unaware that her husband was having an affair with the family governess. Yet another woman had joined the household in January of that year, a Mary Ann Evans, and Chapman's attentions towards his brilliant and freethinking new lodger had filled the house with further tensions and jealousies – common knowledge amongst those who attended the soirées, such as Erasmus. He'd planned to buy and relaunch the ailing *Westminster Review*, people said, and make Mary Ann Evans the sub-editor. Dedicated to progress, it would compete with the

Edinburgh and the *Quarterly*; but something had gone wrong and Mary Ann Evans had mysteriously disappeared, driven out by the other women in the family, some said.[28]

The first week of August was an anxious week for Darwin, despite the interesting diversions of shopping and dioramas and Erasmus's gossip. It passed slowly. More than anything Darwin waited to hear from Lankester, who was not due back from Paris until 7 August. Emma and he had agreed to leave on Saturday, 9 August, so on the seventh he wrote to Lankester, reminding him of their arrangement and telling him that he would send Parslow to Old Burlington Street between 9 o'clock and 10 o'clock on Saturday morning to pick up the manuscript and take it to the printers. This would be leaving it till the very last minute, but this was the whole point of coming to London, after all. He had become obsessed by the care of this manuscript, afraid to leave it for a moment in the care of another until he was sure that it was in the process of printing.

Darwin was afraid. It was only five months since he had left his most precious daughter in the care of Gully and he had lost her. There would be no more risks; but he tried to restrain himself in the letter to Lankester. He didn't want to be demanding or difficult: 'My reason for wishing my own servant to take the MS is that I have not a copy of a page, & I would on no account undergo the labour I have spent on it, & therefore am very unwilling to trust it to the tender mercies of a public conveyance . . . I hope you have much enjoyed the brilliant festivities at Paris.'[29]

So on Saturday morning, with the bags all packed and ready to leave, Parslow set off for Old Burlington Street. He returned an hour or so later with bad news: Lankester was not yet back from Paris. He hadn't even glanced at the manuscript. It had been sitting there on Lankester's desk unseen for nearly two weeks. Darwin was distraught. He could not decide what to do. He cancelled the carriage, resolved to stay another day and send Parslow again in the morning. By the evening, however, he'd decided that he could not make Emma and the children wait for longer. Emma was anxious to be back in Down; they both wanted to see the younger children. Lankester might not be back for several more days. He would just have to trust Lankester, Emma said. So once again Darwin wrote cautiously, careful not to let Lankester sense his fears. He asked the publisher to entrust the delivery to a reliable servant of his own, when the time came, and to send him a note at Down House when the manuscript had reached its destination safely. '*Forgive* my silly particularity,' he wrote, poignantly. He was a man fearful of loss – fearful of blunders. He was a cautious man.

Back in Down, a box of barnacle larvae awaited him in his study with the small pile of letters: more rare specimens from Bate. He sent back the scissors to Newport and wrote to thank Bate. A few days later the first of the printed pages of the second volume began to arrive from the printers – on time. Over the next months these pages would arrive section by section for him to proofread. His labours for the second book were finally over and it looked good, he thought – better than the first volume. He had been right to trust Lankester and to be patient with Sowerby and Bate. His generosity to Bate and his blunders had reaped rich rewards. It was important to keep these friendships and correspondences alive, he told John Lubbock, training him in the ways of a gentleman naturalist. Charm, good nature and good manners – these oiled all the wheels of his complex dependencies on his publishers, his servants, the men who collected for him, and the young collectors whose work he read. It was in his nature to be gentlemanly, to avoid bad feeling, even if sometimes it was a struggle to keep his temper and his patience; but he wasn't entirely sure William had inherited this good nature. He wrote to his son in Rugby with important advice that autumn:

You will surely find that the greatest pleasure in life is in being beloved; & this depends almost more on pleasant manners, than on being kind with grave & gruff manners. You are almost always kind & only want the more easily acquired external appearance. Depend upon it, that the only way to acquire pleasant manners is to try to please everybody you come near, your school fellows, servants & everyone. Do, my own dear Boy, sometimes think over this, for you have plenty of sense & observation.

Love from Mamma. Yours affectionately, C. Darwin.[30]

10

Drawing the Line

*

The Sea of Faith
Was once, too, at the full, and round earth's shore
Lay like the folds of a bright girdle furled.
But now I only hear
Its melancholy, long, withdrawing roar,
Retreating, to the breath
Of the night-wind, down the vast edges drear
And naked shingles of the world.

From Mathew Arnold, 'Dover Beach'

Dawn on 2 October 1851. Darwin, in his black cloak and with his walking stick, slipping out of the house for his usual walk before breakfast, stops to admire the pear trees espaliered along the low brick wall. It is cold. The air is full of the smell of damp leaves, raked into piles along the edge of the meadow; here in the kitchen garden an early-morning mist curls up off the wet orchard grass, and down in the lower fields and in the wood the soil is thick and sticky. Conkers fall in the wood, unheard. Pinioned, clipped and groomed into neat but unnatural shapes, each reaching out for the next in the row, the pear trees look like a row of children's cut-out paper figures, or a row of soldiers on parade in gold and scarlet, branches horizontal like outstretched arms – all different varieties, carefully selected in order to ripen at different times during the autumn and winter. The Marie Louise pear tree, named after Napoleon Bonaparte's empress, in this late-autumn month bears the heaviest burden, branches heavy with gold and russet-coloured leaves sheltering gold and russet-blushing fruit. The leaves are already beginning to turn and fall, but the fruit hangs suspended in air, pendulous, ready to drop. Today Brookes, the gardener, will crop the Marie Louise tree, his calloused hands cradling the fruit's curves as he twists and deftly separates fruit from tree. There may be as many as a hundred pears from this one tree this year, Darwin marvels – more than they have ever had from a single pear tree. He must remember to write and tell William at school, and to tell Brookes to put some aside for William's visit in early November.

They are cousins, these trees, holding hands along the wall – varieties bred by orchardmen in Belgium, France and Holland over hundreds of years, through patient grafting and cross-pollination, selected for his new garden by Darwin's own cousin, the clergyman, William Fox, and carried here as young trees from specialist suppliers. Fox, who had an already established garden in his Cheshire rectory, had chosen pears that would ripen at different times of the year. Fox, the family man, breeder, farmer and clergyman, was now the father of ten, and he still asked after the Down pear trees sometimes in his letters, as if they were family members. Every year these maturing and pinioned trees offered up more and more treasure. After the cropping Etty and George helped wrap the pears in paper, comparing the colours and patterns on the speckled, dappled, veined or spotted skins, smelling or touching them with eyes closed, mouths shaping their French names: Beurre d'Alenbery, yellow with traces of russet; Winter Nelis, rounder, yellowy-green with russet patches; Marie Louise, rich yellow sprinkled and mottled with light russet on the exposed side; Passe Colmar, grass-green with russet spots; and Ne Plus Meuris, green and red and sweet-smelling – cousins with a common ancestor trained to yield at different times, slight but prize-winning variations accumulating generation by generation and continuing to accumulate with each feathering of pollen on to flower or skilled cut of the grafting knife, or pressing of prized seed into warm wet soil.[1]

Hadn't Pliny complained about the taste and texture of pears in Ancient Rome? What poor stock the early wild ancestors of these splendid trees then have been? But since Pliny's disappointed bite, thousands of years of French, British and Flemish skill throughout the cultivated world had bred scores of varieties – thousands of years and long lines of pear-tree descent, each new cross-pollination enhancing shape, taste, colour, smell. Cultivation; and if Pliny were to walk this way, what would he say to the exquisite melting of sweet white nineteenth-century pear flesh? Honey, cinnamon, butter, sugar – tastes nurtured by breeders and orchardmen shaping nature in flowers and kitchen gardens like Darwin's own, improving and modifying without end.[2] What might cultivated pears taste like in three hundred more years with yet greater improvements in agriculture and soil fertilization? Liebig, the German chemist, had been speculating about precisely this in his *Familiar Letters on Chemistry*, the book that lay on his desk, Darwin reflects, that he must remember to post to William: 'Many of our farmers are like the alchemists of old, – they are searching for the miraculous seed, – the means, which, without any further supply of nourishment to a soil scarcely rich enough to be sprinkled with indigenous plants, shall produce crops of grain a hundred-fold.'[3]

And what of the alchemy of parents, training up their own next generation, like gardeners? What of the training he and Emma had given to William, the schooling at Mitcham and now the transplantation to Rugby planned for the boy the following year? Would the decision be regretted later? Would William's young mind benefit from training and pruning or be softened and contracted by it? What about six-year-old George? His time was coming up now. Would a classical education suit such a practical and active boy? He was so good with his hands – so technically minded, such a fine draftsman. Even now he was pestering his mother to agree to the trip to Uncle Francis's house in Etruria, which had been planned for weeks and postponed the day before because of the baby's cold. George needed distracting. Excited by the trip to London and the Great Exhibition in the summer, he was hungry for more visits and adventures, bored and restless at Down, but not yet old enough to begin schooling or tuition. Instead he spent his days drawing ships, soldiers and drummers, and talking about them to anyone who would listen. War was on his mind. War was in the air – the servants gossiped about invasions and the French. Louis Napoleon was dangerous and not to be trusted, they said, a man hungry for power.

While George drew and painted drummers across any piece of blank paper he could find, Darwin made his turns around the

Sand Walk and, during his allotted three hours of work, plunged further and further into the proofs of the second barnacle volume, which had sat on his desk since August. He had just made it past the two-hundredth page.[4] He had decided that, in order to keep life simple, he would not begin work dissecting and writing up the sessile barnacles until all the stalked barnacles were safely in print; there would be no confusing overlaps between books this time. But he had discovered too late and to his enormous frustration that a barnacle that he had named *Xenobalanus siphonicella* in the volume had already been named *Xenobalanus globicipitis* by Japetus Steenstrup, the Danish zoologist, in a previously published paper.[5] It was too late to change the proofs, but he would have to add Corrigenda; the second volume was no longer quite perfect. Furthermore, this would mean that he would have to wait for Steenstrup to send him the publication details of the paper – knowing Steenstrup, this might take months.[6]

Each reading of the barnacle proofs made him less and less certain of his conclusions – what would people make of the *Ibla* and *Scalpellum* males and his conclusions about them? Would these Münchausen stories be believed? Or would the storyteller be laughed at, publicly ridiculed? Every page was the result of rigorous struggle with words and sleepless nights trying to determine how these creatures fitted into recognizable genera; blood and sweat in every phrase. But how many people would actually read them? What mark would they make on the world? The Ray Society would send out the volumes to its members, and it would send him a further twenty complimentary copies to distribute himself, but he would have to seek out reviewers for himself, using all his arts of persuasion. He would have to publicize his barnacles for himself.[7]

Hooker was now going through his own classification nightmares in the gardens of Kew. After all the glamour of Paris and the jurors' reports of the Great Exhibition he was now trying to analyse his own Indian botanical hoard, his roomfuls of carefully labelled boxes of dried leaves and flowers, now housed in Kew's Temple of the Sun and in a shed behind the Orangery. He was not alone: he had a partner in

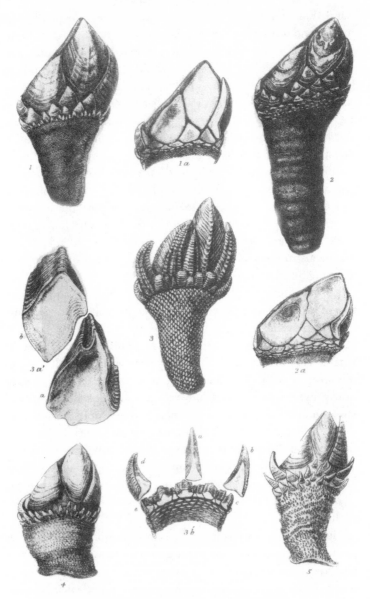

24 The Outside of a Stalked Barnacle

his labours, the fellow botanist, traveller and collector, Thomas Thomson. Their joint collections amounted to 150,000 specimens and six or seven thousand species. Hooker's first problem, just as Darwin's had been, was to find a way of making sense of the confusion of names that had been applied to the plant species he had collected by the various botanists and amateur plant collectors who had published papers – synonym after synonym, some correctly applied, some incorrect, like patchwork. He and Thomson both, like Darwin, deplored 'the prevailing tendency to exaggerate the number of species and to separate accidental forms by trifling characters'.[8] The classification of these plants, like the classification of barnacles, was in chaos as a result of collectors splitting and resplitting groups and species. He and Thomson had to forge some degree of order in their naming strategies before they could even begin to work out patterns of geographical distribution and diversification. They set out to reduce the numbers of species. What was needed was a degree of inclusivity – clear boundaries between species listed as large groups by patterns of affinity.[9]

The reality was much messier altogether. Hooker was now facing many of the impossible questions and impassable obstacles that Darwin had himself faced as he began the barnacle work, trying to define minute variations between one group and another. Where did one group begin and another end? Where were the boundaries? What were the critical parts of an *Ibla* or an orchid that made it *essentially* an *Ibla* or an orchid? And which plant or animal *part* served as the absolute measure of difference? The particular problem that faced Hooker in November 1851 was the New Zealand flowering plants, which showed, like Darwin's barnacles, a terrifying degree of variation. He wrote to tell Darwin of his problems, knowing that these questions pressed on his friend's species theories: 'There are 4 or 5 genera of flowering plants in N.Z. so large (for an Island where all the genera are small) & so disgustingly protean, that I am again reconsidering old Bory de Vincents dictum as to the variability of Insula species . . . Coprosma, is almost peculiar to N.Zeald & for the life of me I do not know how to draw the line between there being only one species, or 28!'[10]

How to draw the line indeed? And to which side of the critical line would Hooker's frustrating classification forays take him when he came to publish? He had written from the Himalayas that, try as he might, he could not see sufficient evidence of Darwin's species theory in the flowering plants of India; but gratifyingly, Darwin noted, he was at least prepared to look at his botanical specimens through the lens of the species theory. He had an open mind, and he had been impatient with Darwin, wanting him to return to the species theory, to finish it, send it out into the world. He was bored with the barnacles. He knew what explosive material Darwin had tucked away in his drawers and he wanted his friend to attend to it, gather the evidence, make the case, so that other people could use it.

With Hooker, the secret sharer, back in England and writing lines like these, Darwin felt the pressure of the formed but not yet ripe theory more heavily than before – felt its weight upon him. He did not write about it to Hooker, yet it lay between them like a spectre of expectation, and he felt compelled to allude to it in each letter he wrote to Hooker on his return – a gesture, a glance, a promise that it had not been forgotten. Yet he had to finish the barnacles, and the enormous and 'disgustingly protean' community of sessile barnacles still lay before him, unmapped and unplaced, filling him with dread.[11]

It was late in 1851. The two volumes of stalked barnacle classification had already taken him five years and perhaps broken his health. How much more time would the sessiles take – the most difficult task of all? He had written out a plan of work that made him feel better, a scheme that should take only a few months. He had made a list of all the sessile barnacle groups, saving the trickiest to the end: the group of truly aberrant burrowing barnacles would be his *pièce de résistance*, his swansong, before he passed on to the tiny group of fossil sessile barnacles and had done with barnacles for ever. Mr Arthrobalanus and his fellow deviants lay at the end of that plan of work – still waiting in the wings. When he reached him, the labours of this latter-day Hercules would be almost at an end.

Sulivan came for dinner in December 1851. Lieutenant Sulivan of the *Beagle*, big-hearted, impatient with himself and with everyone around him, wanting to get on, wanting to get out, pacing the deck, was now Admiral Sulivan. Sulivan the hunter, the restless, the friend with whom Darwin had stirred plum pudding in an empty rectory on an abandoned island one Christmas Day, now in his forties, too, and with half a dozen children, brought his restlessness, his sleeplessness and his hunger to the quiet of Down in December. Over dinner he stirred them all up with his stories of the Sullivan family's adventures and his predictions of impending invasion. George was besotted.

Disembarked from the *Beagle*, Sulivan, like Darwin, had married and then taken his bride, a steely and adventurous admiral's daughter, to sea again, as part of his commission to survey and map the inhospitable Falkland Islands. So while Charles and Emma had sown their orchards and borders at Down, Captain and Mrs Sulivan had sailed the coasts of the Falkland Islands. Without chloroform, Mrs Sulivan had given birth to babies on ship and on land. Falkland, their eldest boy, had been, they claimed, the first European to be born on the Islands. Then, while Sulivan petitioned unsuccessfully for a command back in Plymouth, his health suffering, they had decided to try their luck back out in the Falklands, not this time as surveyors, but as breeders of horses and sheep. Sulivan told the assembled guests how, granted three years' leave of absence, he had returned from the Admiralty on a Monday morning to say to his wife, 'Will you be ready to leave by Thursday?' to which she had replied, 'No, but I will be by Monday next.'

The family had sailed for the Islands with a recently hired governess, five children under the age of eleven, a maidservant, a manservant and his wife, six thoroughbred horses for breeding, sheep and a piano. By the time they arrived there were several men at the docks, waiting to propose to the maidservant, for there were virtually no unmarried women on the island. The maid was married within weeks, Mrs Sulivan told the Darwins, leaving her to manage her new house and five children almost entirely unaided.

Sulivan had brought Darwin a present: not snuff or port for the cellar, but a crustacean taken from the stomach of a dolphin, drowned in the Falklands kelp. He knew how to please his old shipmate. It was, Darwin told him as he examined its curled horns and carapace, a very rare specimen indeed, so rare it must be seen by the eyes of James Dana in New York, who was completing his two-volume life's work on crustacea. So he slipped it, carefully nested in a wooden pill box, inside the parcel containing two complimentary copies of his barnacle volumes he was sending Dana. Dana never received the tiny, carefully wrapped wooden box. Officials at the New York Customs House opened Darwin's parcel and lost the box with its well-travelled and rare inhabitant somewhere amongst the piles of letters, parcels, brown paper and string.[12]

Darwin found Sulivan's emigration tales alarming. This was hardly the kind of grand future he had fantasized about for William and George; but then Sulivan, in choosing the Falkland Islands, had hardly chosen the land of milk and honey. Even the journey home had been terrifying. Sailing for England – not with five children but with six, for Mrs Sulivan had given birth to another baby during their last months in the Falklands – the ship on which the family sailed had been too heavily laden. Two members of the crew had stoked up a mutiny within weeks of embarking on the ninety-day voyage. However, when the mutinous crew had gone below to make their deliberations, quick-thinking Sulivan, the Captain and the mate had nailed down the hatches, lowered the sails and sailed the ship themselves for the few days it took to starve the crew into submission. Mrs Sulivan, who had nursed her children through sieges in Montevideo and the wild climatic changes of the Falklands, was uncomplaining. Perhaps, then, Charles and Emma were better off here on this quiet ship on the Downs, safe from pirates, mutineers and war.

Sulivan brought anxieties and sleepless nights, too, to the quiet world of Down. He was a man of war. Son of a naval commander, nephew of three naval officers, his return to England from the Falklands had been prompted not a little by the smell of war. After the women had retired from table that night in December, he reminded the country gentlemen assembled around Darwin's table that Napoleon was dangerous. The coup d'état the French leader had just pulled off in Paris was likely to be only the beginning. He was ambitious; he had plans; he wanted an empire – called himself an emperor. Moreover, the French were clever, unscrupulous and deadly enemies. If you were Napoleon, he challenged his guests, flushed by wine, where would you enter England if you were determined upon conquest? Where would be the easiest point of entry? The Kent coast, he told them triumphantly, drawing maps and diagrams with his fingers on the tablecloth, was the easiest place from which to march on London; and sleepy villages like Down lay right in the Frenchman's path. Imagine a French battalion surrounding your house at night, he challenged them, pushing the salt cellars with sinister precision further toward the fruit bowl piled high with winter pears. How would their servants fare? Would they fight? When was the last time they handled guns? What kind of men did they have about them? How safe would their womenfolk be?

What the country needed, Sullivan intoned, was a volunteer corps for the south-eastern counties, Martello towers in all the coastal villages, good road connections. Every hunting district should form troops of well-mounted young men armed with swords and light rifles. The men around Darwin's table that evening urged Sulivan to write letters to the *Naval and Military Gazette*; some of the *Gazette*'s readers might have government influence, they reasoned. Darwin wondered about the state of his own guns and how men like Parslow might react to an invasion. There were certainly plenty of young men in the village who would fight, but would they be ready?

Darwin worried his way through Christmas and into spring. 'My nights are *always* bad,' he wrote to his cousin Fox in March, and at night his worries magnified to monstrous proportions; his thoughts twisted and turned through worries about money, the future of his children, inherited weaknesses resurfacing in his children (he and Emma had started to keep a diary of five-year-old Lizzie's strange twitches and pronunciation habits, worrying that she might have an inherited defect) and the French invading Down. 'My three Bugbears are Californian & Australian Gold, beggaring me by making my money on mortgage nothing – The French coming by the Westerham and Sevenoaks roads, & therefore enclosing Down – and thirdly Professions for my Boys.'[13] He continued: 'I congratulate & condole you on your tenth child; but please to observe when

I have a 10th, send only condolences to me. We now have seven children, all well Thank God, as well as their mother; of these 7, five are Boys; & my Father used to say that it was certain, that a Boy gave as much trouble as three girls, so that bona fide we have 17 children ...'

He had capitulated on William's education, agreeing to send him to Rugby rather than the experimental schools they had considered at first; but his mind was not at rest about it. How would Rugby shape him? The Lubbock boy, John, now eighteen and still riding over from High Elms to Down House on his pony to work alongside Darwin at the microscope, had turned out well enough after his Eton education. He was clever and curious and he could write well and precisely. Being taught to write well – that was perhaps the most important training of all, the kind of training his tutor, the Revd Henslow, had insisted on at Cambridge. He had even tried to persuade Henslow to write a little training book for children: 'I often reflect over your inimitably (as it appears to me) good plan of teaching correct, concise language & accurate observation, namely by making your pupils describe leaves &c. . . . a most useful volume might be published . . . What a habit it would give to youths of thinking of the meaning of words & what powers of expressing themselves! Compare such habits with that of making wretched Latin verses ...'[14] Correct and concise language still eluded Darwin in his daily struggles to put barnacles into words.

His health kept him living as a hermit, he told his cousin Fox, and if this had not been so, he would indeed have been off to Yorkshire to see and attempt to describe the effects of the bursting in February 1852 of the Holmfirth Reservoir, a rare opportunity to witness the effects of violent deluge acting upon a landscape. The mill owners, in an attempt to control the erratic water supplies of the region, had built the reservoir, a dam of some 340 feet in length and ninety-eight feet high with a mass of earth on either side, above a valley of mills, villages and shops with a large population of mill workers and shopkeepers. Having successfully controlled the water flow and supply to the factories, the mill owners, engineers and commissioners had subsequently neglected the maintenance of the dam and the whole structure had quickly become rotten and defective. Heavy

winter rain had choked the valves of the overflow and on the afternoon of 5 February the embankment had burst and floodwaters
swept through the hamlet of Holm like a tidal wave, carrying everything before it: churches, shops, barns, and cottages. Hundreds of
inhabitants had drowned, often whole families together.

Darwin, so long a student of the effects of natural disasters such as
volcanoes, earthquakes and tidal waves, wanted to see the shape of the
flooded valley, in the aftermath of this catastrophe, with his own eyes.
The papers described the effects, but journalists were inevitably more
interested in the tragic human consequences than the geological ones:

When morning broke, the spectacle presented by the once busy valley was
fearful in the extreme. The shattered fragments of walls retained the ruins of
other buildings or their own. – Mill wheels, timber, roofs of houses, fractured carts, pieces of cloth, and household furniture were intermingled with
huge rocks, or half buried in stones and mud. The boilers of steam-engines
loomed large in the bed of the stream or stranded in the gardens – while
here and there a drowned corpse was to be seen lifeless on the water-left
shingle, or buried in ruins; to the horror of this spectacle was added the
presence of numerous skulls and other human bones which the torrent had
washed out from the graveyards.[15]

On 26 March 1851, almost a year to the day after he had taken
Annie to Malvern, Darwin began reading a book by a promising
young naturalist who had been making a name for himself in micro-
scopical studies and marine invertebrates, Philip Gosse's *A
Naturalist's Sojourn in Jamaica*. Gosse, like Huxley, was one of the
new voices in zoology: he had already published a number of articles
in the *Annals and Magazine of Natural History*; but although his
prose style was lively and elegant like Huxley's, and his observations
acute, he was publishing his books at this time principally for the
Society for the Promotion of Christian Knowledge and, as a zealous
Christian, his writing tended to be descriptive and worshipful rather
than theoretical or speculative. He had returned from his travels to
Canada and Jamaica like Darwin and Huxley, with copious zoological
and botanical notes, and now he was seeking to invest this knowledge
in publications that might secure him an income.

Darwin found himself sharing many of Gosse's ideas and passions. His Preface to *A Naturalist's Sojourn* was brave and assertive for a man entering the word of zoology. He had begun this first major book with a clarion call, challenging zoologists to abandon the study of dead and dried things and instead commit themselves to the study of live creatures. He was even now, people said, working on finding a chemical formula to produce artificial seawater so that naturalists could study live sea creatures in glass tanks instead of just dissecting them. This was a different zoology – a humane one, admittedly, but a world away from the philosophical problems of anatomical structure and shape-shifting that so taxed Darwin. Gosse was interested only in the here and the now of the creatures he studied, not in how they had come to be.[16]

Joseph Hooker and his wife Frances came to stay at Down in April, the week in which the family would pass the first anniversary of Annie's death. It would be a distraction for them all, for the children loved their father's friend, and it gave Darwin an opportunity to cross-question Hooker on the problems of classification now facing him. Frances watched Joseph playing with the Darwin children, down on the floor throwing pillows, playing at bears, imagining herself at the centre of such a family, a family that they could not begin until Joseph had secured himself a position, she told Emma. On the anniversary of Annie's death, Emma, grieving and melancholy, suggested that they all walk to the top field to fly George's new kite, but in high winds the new kite tangled in the tall trees, tearing its coloured silk into ribbons. Emma was resourceful as always, the Hookers marvelled: while she comforted George, Parslow climbed the tree to rescue the tattered kite. She would mend it, she reassured George. All would be well: the kite would fly again.[17]

Hooker had now completed much of the classification of the New Zealand plants and had begun to think about writing an introductory essay that would deal with the philosophical problems facing the botanist. Darwin was impatient to know how Hooker would conclude, how he would represent these insurmountable problems but he reminded Hooker that he must say nothing of the species theory in this introduction – not yet.[18] He was sworn to secrecy. Darwin was excited, Hooker exasperated. How could he keep natural selection out of the introduction when it explained so many of the variations in New Zealand flora? There were other obstacles to publication, too, for a philosophical botanist such as himself: no one could even begin to speculate on the species theory, he complained, until the competing naming methods settled into a common system that would at least mean that zoologists and comparative anatomists were speaking the same language, not this current tower of Babel dominated by ignorant noodles who couldn't tell a cabbage from a cabbage palm.[19]

During Hooker's visit the two men spent long hours discussing the implications of Hooker's difficulties with classifying the flora from New Zealand, after breakfast in the study, or later in the day, strolling round the Sand Walk, Etty or George in tow. Pumping, Darwin called it; he was pumping Hooker, drawing upon his knowledge, forcing him to confront certain theoretical problems.[20] As their shared questions settled into shape, Darwin formulated a list of seven points for discussion and wrote them out as a memorandum for Hooker to take away with him: 'Questions for Hooker', he titled it. They were thinking through the same set of questions about the variability and distribution of species, with Hooker representing the Vegetable Kingdom and Darwin representing the Animal Kingdom. It was a rare event. The questions formulated by Darwin at Down, annotated and answered where possible by Hooker, passed backwards and forwards between them for months. These were exciting times: anything was possible. This secret sharer, mountain climber, rhododendron hunter, border crosser, man of courage – would he stand by the species theory when it came under attack?

Now that the *Ibla* and *Scalpellum* stories were out, Darwin waited with bated breath to hear whether they would be believed by the subscribers to the Ray Society. In addition he had sent out almost all his twenty complimentary copies to the most important international zoologists working on marine matters, and awaited their verdict. Letters began to arrive slowly, but the approval was strong and unwavering. He *was* being believed, despite the hesitating and provisional nature of his conclusions about the complemental males. Johannes Peter Müller wrote from Berlin, then James Dana from

America. Their congratulations gave him the heart to go on, he replied. Then Professor Richard Owen wrote in July with approval that was essential if the books and their theories were to be well received in London circles. Darwin was gratified and grateful for the endorsement, though he was happy to confess that he could see weaknesses in his accounts of the living stalked barnacles: he had not given over enough time to the anatomical part, he feared, because he had been anxious about time. He was none the less delighted by Owen's letter: 'Pray believe me in a great state of triumph, pride vanity, conceit &c &c &c, Yours sincerely, Charles Darwin.'[21]

Whilst he had thought that 'the only part worth looking at [was] on the sexes of *Ibla* and *Scalpellum*',[22] he was being commended for having discovered the homologies of the shell and the external part of cirripedes as well, discoveries that had been long awaited by comparative anatomists. In the late 1830s the French comparative anatomist, Henri Milne-Edwards, had worked out that an archetypal crustacean has twenty-one segments or body parts. When Vaughan Thompson discovered the metamorphosis of the barnacle from free-swimming larva to fixed adult at about the same time, the way was clear for a careful observer to map the barnacle anatomy against the twenty-one segments of Milne-Edward's archetypal crustacean. This was one of the tasks Darwin had undertaken from the start, but was only part-way through by the publication of the 1851 volumes, for although he had been able to find seventeen of the crustacean segments represented in barnacle bodies, four were missing. This meant that he had a method by which to compare body parts – it gave him a basic plan against which or over which he could plot his 'deviants', like Mr Arthrobalanus, in order to see how and where they had diverged from the crustacean body plan. Then, of course, he could speculate about what evolutionary advantage this deviation might have given them. Mutation of body parts meant that the living creature could accommodate new functions: what might once have been a pair of limbs for walking might elsewhere have transformed into cirri for feeding for instance. Other organs, such as the missing abdominal segments, might simply have atrophied over millions of years of non-use. Nature had simply rearranged the body parts of an archetypal crustacean over thousands of years to fit it for new conditions.

Sleepless and racked by toothache in June, Darwin read Dr Henry Holland's new book *Chapters on Mental Physiology*. Holland, too, was taking risks and being tentative about them – these were notes, he insisted in his introduction, not a complete treatise and the 'topics treated of are such in their nature as perpetually to bring us to the very confines of metaphysical speculation'. He wanted to explore the influence of the mind over the body, particularly the bearings of mental action upon 'morbid disease' either 'directly or indirectly – as cause or as effect'.[23] Unlike many of his peers, Holland was also a border crosser, like Darwin, convinced that there were no absolute lines to be drawn between mind and body, nor between certain mental states such as sleep and wakefulness: 'Take . . . a person seeking rest on an easy bed, or under the influence of pain or disordered digestion. Obviously to himself, as those around him, there is an incessant alteration of state, testified by various bodily movements, by partial consciousness of external objects, by dreams broken and renewed – a strange interlacing of the two conditions, which thus divide our existence.'[24]

However, whilst Holland was clearly excited by new biological ideas and medical practices, and accepted the flux of nature and of mental states, even showing his understanding of the debates about the mutability of species, he concluded that there simply was not enough evidence so far to prove any of these ideas as anything more than speculations.[25] Two days later Darwin experienced his first deep and unfluctuating sleep in months: his London dentist, Mr Waite, put him under chloroform so that he could extract five teeth.[26] It cost Darwin a guinea, he wrote to Fox, but it was worth it for 'this wonderful Substance'.

Above all, he was tired – tired of the sleepless nights, of the Water Cure regime, of the barnacles more than anything, though they had still not finished tormenting him, he wrote to Fox in October: 'I am at work on the second vol of the Cirripedia, of which creatures I am wonderfully tired: I hate a Barnacle as no man ever did before, not even a Sailor in a slow-sailing ship.'[27]

His ship – his barnacle-encrusted volumes – was excruciatingly slow-sailing. This time he was determined not to be further slowed down by a tardy illustrator. He had employed another Sowerby for the illustrations for the third volume, George Brettingham Sowerby Jr, the nephew of James Sowerby, who had been so frustratingly slow. George was less of an artist but more efficient. Darwin had even invited him to Down House for a two-week visit so that they could work together on the drawings for the third volume. These were Herculean labours, and yet the most difficult of all the barnacle territories still lay ahead. Turbulent seas: something had to yield. A few months before, on 22 August, he had decided to give up the Water Cure regime, which he had only been continuing in a rather intermittent way since Annie's death the previous spring. It did not seem to be bringing the results that it had done the previous year, and as winter was not so far off, it seemed a good time to give up the cold torture.[28] His health, mysteriously, improved quite dramatically throughout the last months of the year, giving him renewed vigour to face the last deviant barnacles.

John Lubbock was a welcome presence during these labours, working alongside Darwin in his study for a few hours every few days.[29] The eighteeen-year-old had been working on crustacea most of the year with his new microscope, and now Darwin had given him the task of dissecting and drawing water fleas, a transparent crustacean called *Daphnia*, which swarmed in Down Pond. In November John began a new natural history notebook recording: 'Took my work up to Mr Darwin and after coming home copied out neat a good deal of what I could say about it.'[30] The water fleas, like Darwin's *Ibla* and *Scalpellum*, had inventive reproductive patterns: they were almost exclusively female, reproducing by asexual methods; but occasionally, when conditions were poor owing to a lack of food, low oxygen supply, a high population density, or low temperatures, they would produce males and reproduce sexually. John Lubbock was using a camera obscura up at High Elms so that he could draw infusoria from nature, and his drawings were becoming remarkably accurate, Darwin noted with pleasure, all the more aware of the limits of his own drawing skills. John would be ready to publish some of his discoveries on Darwin's crustacean collection soon, particularly his discovery of new species of *Labidocera* and *Calanidae*.[31] Darwin was ambitious for his young friend and was keen to help his career as a naturalist. When James Dana wrote to tell him that he had completed his three-year crustacean research, Darwin wrote to congratulate Dana but also to commend his young friend: 'I have a neighbour, who is very anxious to see this work; he is the son (very young) of Sir J. Lubbock, the great astronomer & Banker, who has taken up the smaller Crustacea with great zeal, & will soon publish a paper on a subgenus of Portia.'[32] He would now write to the editors of the *Annals of the Magazine of Natural History* on John's behalf.

Darwin had still to return to the problem of how and why barnacles had evolved burrowing powers – why had these groups, found in very different places around the world, developed the ability to make their homes in the bodies of sharks or in the cavities of rocks, rather than secreting their own houses around them like other sessile barnacles? How did they fit into the barnacle world? It was now time for Mr Arthrobalanus, *Verruca* and *Alcippe*, to step back on to the microscope stage and time, too, to renew correspondence with the Newcastle naturalist, Albany Hancock. The two of them had continued to correspond now for some three years about the burrowing powers of barnacles. Darwin had even written to Hancock a long and frustrated list of problems on Boxing Day of 1849, and here he was again with a similar set of questions covering several pages written on Christmas Day of 1852.[33] This was the last sheer cliff face before the summit, and Hancock would be his fellow climber. He had set out all those years ago, he explained in his Christmas letter to Hancock, convinced, like Hancock, that *Verruca* had *mechanical* burrowing powers; it did its own digging somehow. But having found three other burrowing barnacles around the world, he had come to believe that burrowing barnacles made their dugouts through chemical action. He was the first to admit that this theory was still hypothetical. Now he must find evidence. The last entry in his journal for 1852 reads: 'Began *Verruca*'.

Hancock was delighted to be a co-sailor on this last voyage of discovery, now that Darwin was going to attend to *his* creature, the barnacle he had discovered in 1849, the burrower *Alcippe lampas*, and particularly now that he had read and admired the second of the barnacle volumes. He wrote to express his admiration for the book and Darwin replied: 'I am quite delighted at what you say about my little friends, the complemental males; I greatly feared that no one wd believe in them; & now I know that Owen, Dana & yourself are believers, I am most heartily content.'[34]

Other believers would be press-ganged on to the burrowing-barnacle ship, for he needed more burrowing specimens. Favours would have to be called in. Charles Spence Bate, the dentist who lived by the sea in Plymouth, was recruited to collect *Verruca* specimens from both calcareous and non-calcareous rocks so that Darwin could determine how they burrowed. Bate spent cold mornings at low tide down on the Devonshire beach scraping *Verruca* off different kinds of rocks, preserving them according to Darwin's meticulous instructions and labelling them, all before he started a day's work pulling teeth.

Only a week into the new year, Darwin's spell of good health broke into bouts of flatulence and vomiting, perhaps exacerbated by the rich Christmas food. Now the grotesque and highly magnified body parts of *Verruca*, *Alcippe*, *Arthrobalanus* and *Proteolepas* coloured and streaked and twitched their way across Darwin's dreams. When he closed his eyes, he could see them all in their puzzling nakedness, working to make a cavity for themselves. These were barnacles with the strangest anatomies of all – *Alcippe* was 'one of the most difficult creatures', he told Hancock, who had discovered it in 1849, and he simply did not know how to classify it. It was peculiar in almost *every* body part and, strangest of all, it had no rectum or anus, so that it had to '*always* eject . . . its excrement from its mouth'.[35] When he wrote to Fox with news of the birth of Hooker's first child, he couldn't resist telling him about the vomiting, anus-less *Alcippe*: 'I have this morning been dissecting a most abnormal cirripede, which after a good meal has to vomit forth the residuum, for there is no other exit! 'I heard yesterday from Dr Hooker, who married Henslow's eldest daughter, of the birth of a son under Chloroform, at Hitcham.'[36]

Embedded in the flesh of this vomiting female monster, with her vestige of a rudimentary and atrophied penis, were clusters of males so strange that he didn't even recognize them as barnacles at first. Even now that he had seen the miraculous diversity that barnacles were capable of, he was still fooled by these, and had thrown the males away in the first dissections, thinking they were parasites of the *Bryozoa*; but looking again, he wrote to Hancock, he could see that these were more minute complemental males, like the *Ibla* males, living in the flesh of a large female, engorged with sperm:

The male is as transparent as glass . . . In the lower part we have an eye, & great testis & vesicula seminalis: in the capitulum we have nothing but a tremendously long penis coiled up & which can be exserted. There is no mouth no stomach no cirri, no proper thorax! The whole animal is reduced to an envelope (homologically consisting of 3 first segments of head) containing the testes, vesicula, & penis. In male Ibla, we have hardly any cirri or thorax; in some male Scalpellums no mouth; here both negatives are united . . . I believe the males occur on every female: in one case I found 12 males & two pupae on point of metamorphosis permanently attached by cement to one female! [37]

When he had discovered the *Ibla* males, he had assumed them to be unique, but now he had found several examples of the emergence of separate sexes, all in different corners of the barnacle world. It was a rare adaptation but a singularly important one, for it helped to show further evidence for his developing ideas about sexual selection. When reproduction was essential for the survival of the species, having the longest penis in the animal kingdom was sometimes not enough. Supplementary males were an insurance strategy for less risky reproduction, by providing the species with a range of diverse reproductive methods.

January had been unusually warm but as February began, the barometer fell alarmingly and soon the country was beset by gales and snowstorms. Snow fell almost every day. George, now seven, took his ice-slide down to the frozen village pond to join his friends every morning and came back with his face flushed and hands blue. Emma had nursed William, Etty and George through an attack of mumps in early February, and by the beginning of March four-year-old

Francis and five-year-old Lizzie had the tell-tale swollen faces and sore throats. William's infection had delayed his return to school and he gratefully holed up in the warmth of the schoolroom with the copy of the *Life of Napoleon Bonaparte*, which his father loaned him. Emma helped him pack the book next to his ice-skates when the day came to return to school.[38] There were icicles along the eaves and some of the late winter-ripening pears blushed pink and russet under thick white coats.

Darwin was delighted to have abandoned his daily douche back in August, for it would have been almost impossible to pump the water in this weather, let alone to tolerate icy water on naked flesh. The newspapers were full of accounts of accidents caused by exposure: three soldiers on Dartmoor had wandered into a snowdrift; the prisoners and guards of Dartmoor Prison had been virtually cut off from the rest of the world and were half-starved; railway tracks were blocked, and numerous ships had foundered in heavy gales.[39] The weather conditions had produced other meteorological spectacles and dramas: sightings of the aurora borealis across England and Scotland and a mysterious fireball that had nearly destroyed Lincoln Cathedral on 23 February, hitting the tower and exploding in gale-force conditions and heavy snow. Three days later a hurricane hit the west coast and swept across the country, wrecking ships in the ports and out at sea, pulling off masts, rudders, anchors and sails. Hundreds of boats were driven ashore in thick snow.[40]

Darwin's collectors, Hancock and his friends on the Northumbrian shores and Bate down on the Devonshire coast, persisted, despite snow, ice, driving winds and mountainous waves. Postmen persisted, struggling through snowdrifts. Railway workers persisted, clearing railway tracks of snow. Barnacles continued to arrive at Down throughout February, and Darwin continued to write.

With the study fire banked higher than usual, Darwin struggled to make out the *Alcippe*, hoping that here he had found the closest kin for Mr Arthrobalanus, for the two species had several features in common: they both burrowed, they had approximately the same number of limbs, arranged in similar patterns; but *Alcippe* was difficult to link to any other kind of barnacle, almost impossible to compare because of the scale of its deviation from the barnacle archetype. He could not decide how to place the species and Hancock could not offer any help when Darwin wrote to ask for it with desperation: 'My surmises are too vague & too long to tell in this note, & perhaps all of a blunder, but I am dreadfully perplexed,' he wrote, worriedly.[41] Hancock wrote to assure Darwin that he had confidence in the direction that Darwin's conclusions were taking him, 'but my knowledge of the Class is so imperfect that I have no great confidence in my own opinion on the subject. I am therefore well pleased that this curious animal is now in such competent hands, and have no doubt that you will find for it its proper place in the Classification, how difficult soever the task.'[42] Darwin was on his own. By now no one knew the order as well as he did. No one could make this decision with him.

As he prevaricated, a letter arrived from Edwin Lankester from the Ray Society asking – politely but with some insistence – when the third volume was due to arrive. Darwin replied desperately, 'I have at least 6 weeks of dissection to do. Before going to press, I must have a few weeks rest, & I do not think I shall be able to send you my M.S. till the beginning of August: but I will & can make no other pledge, except that I will work every day without exception, on which I can.'[43]

Now he had to turn his mind to Mr Arthrobalanus. It had been eighteen years and two months since his first encounter with this monster in its conch shell on the Chilean beach – nearly as long as the entire life span of the young John Lubbock, who had turned eighteen the year before. Nearly seven years ago he had thought that a month would solve the questions Mr Arthrobalanus's aberrant body raised. Where had the time slipped away to? Would all the knowledge he had accumulated in those seven long years equip him any better for understanding Mr Arthrobalanus?

11

Manoeuvres and Skirmishes

*

Do not despise the creatures
because they are minute . . .
doubt not that in these tiny
creatures are mysteries more
than we shall ever fathom

Charles Kingsley, *Glaucus or the*
Wonders of the Shore, 1855

Early March 1853. The snowstorms abate for long enough for a few snowdrops to flower and the village pond to thaw, but on 15 March the winter winds and snows return with great severity. The Darwin children have tired of playing snow angels, skating and sliding, tired of the white light and picturesque snow-muffled fields, and the Down servants complain about the incessant extra work the snows have brought. The fields are covered again. The children, confined to the house, are listless and bored. Everything is more difficult. The rooms feel cold and draughty even with the fires banked up. Parslow polishes snow-stained or mud-encrusted shoes, digs out the drive, removes the icicles that have grown dangerously long. Everywhere the light is intense and shadows hard and sharp – Skimmed Milk White, Blue White, ghostly white – but at dawn, when Darwin walks the Sand Walk, the rising sun stains the snow a soft pink: Aurora Red.

Darwin holds a glass specimen bottle up to the study window, blows off the dust and shakes it slightly so that the yellow-flecked fluid swirls angrily like a tiny whirlpool held within his hand. Beyond the window, a fresh snowstorm breaks, wind-blown across the driveway; the sky is swollen dark grey washed thinly with carmine. The label on the bottle in his hand has yellowed and the brown ink faded, but he can still make out the words written in the cabin of the *Beagle*: *Balanidae*; Mr Arthrobalanus. 'Our little fellow,' he and Hooker used to call him – our little *invisible* fellow – but under the microscope such a grand and curious monster. *Mr* Arthrobalanus. He liked the name. He hadn't quite been able to adopt the Latin name he and Hooker had invented: *Cryptophialus minutus* – too many syllables; and if he, Charles Darwin, unique in all the world for his knowledge of barnacle anatomy, the man whose eyes had seen and described more barnacles than any person living or dead – if he could not puzzle out this anomaly, then no one could. He had two weeks to do it – two weeks according to the tight barnacle schedule he had reworked since Lankester's pressing letter.

There must be about eight or nine specimens still floating in the jar, he estimates – enough to make up several slides. He will need separate specimens to produce a slide each for the cirri, pupae, reproductive parts, stomach and mouth. He gathers his materials: the boxes of glass slides and glass cover slips, the jar of asphalt for sealing, the caustic potash for dissolving the shell, the microscope and micrometer, the notes and papers and drawings, the micro-scope slide catalogue, which now lists 254 separate slides, a sheaf of fresh paper and a jar of cobalt-blue ink.[1] All around him hundreds of barnacle valves and body parts, some as ancient as dinosaur bones, sit in their labelled coffin-like pill boxes, waiting to be returned to their owners or posted to the British Museum. Philip Gosse had complained in his last book that zoology was too much like a necrology – a science of dead things: dry skins, furred and feathered, blackened, shrivelled and hay-stuffed, bleached and shrunken, suspended by threads and immersed in spirit in glass bottles. Zoologists should study their creatures alive and going about their business, he said, in order to really understand them.[2] Mr Arthrobalanus, though, was almost invisible to the naked eye and almost invisible within his conch-shell home. There was nothing to see, and its barnacle business all went on down there, invisibly, in the hole of the shell. Mr Arthrobalanus didn't walk, sing, fly, migrate or change his colours. The structure and layout of his body was his singular miracle and this was only made visible by using acid to dissolve the shell, dissecting pins to open up his body and a microscope to enlarge it.

Arthrobalanus was distinctly deviant, Darwin remembered, leafing through the pages of notes gathered in his files, the drawings he and Hooker had put together at Kew, his own notes from the *Beagle* and the final paper he had written about the barnacle and sent to Richard Owen for comments, still unpublished because still incomplete. Back *then*, in 1846, scribbling his first microscope notes in the *Beagle* cabin, he had thought Mr Arthrobalanus unique – the only naked burrowing barnacle ever discovered; but since then his collectors had brought him other burrowing barnacle deviants, *Alcippe*, *Verruca* and *Proteolepas*. He remembered his amazement at discovering that one of the Arthrobalanus specimens he had dissected had what seemed to be a double penis, but no ovisac, which seemed to suggest it was not hermaphrodite but fully male. Now that he had found fellow deviants from the barnacle hermaphrodite norm – the spectacular *Ibla* and *Scalpellum* complemental males – he was able to return to Arthrobalanus with different expectations and a more open mind about their reproductive arrangements. When he looked again he discovered now that *Mr* Arthrobalanus was actually *female*. What he had assumed to be a double penis was part of the cirri, and, as in *Ibla* and *Scapellums*, the males lived parasitically upon the much larger females, mere bags of spermatazoa embedded in female flesh. Mr Arthrobalanus was now an 'it', with male and female parts separated out yet mutually interdependent. Nevertheless, Mr Arthrobalanus had assumed a male identity since Darwin's first presumptions and would continue to be imagined and referred to as male in Darwin's correspondence.

What more could he find here? What more could he say? He needed to determine Mr Arthrobalanus's place amongst fellow barnacles, find a family for him; and what would Mr Arthrobalanus's presumed relation to *Alcippe* finally be – if there was a relation at all? In all the world *Alcippe*, fellow burrower, was most likely to be his nearest kith and kin. This was why he had placed them together in the barnacle list and dissection schedule.

First Darwin used his pins to dissect Arthrobalanus's impossibly minute egg-shaped larvae, teasing out developing adult parts, searching even inside the anterior horns, in which he found prehensile antennae, every part perfect. This is where the secrets began – in the larvae. There, pleasingly, were the swimming legs that corresponded to the second, third and fourth thoracic limbs of the crustacean archetype – as he had expected. All was going to plan; but as he examined more larva specimens in different stages of maturity over the next few days, he was astonished to find that Arthrobalanus metamorphosis was utterly *unlike* that of the *Alcippe*. Although barnacle young were generally free-swimming, the young Arthrobalanus larvae never developed swimming legs, and could only crawl about using their antennae. Structurally, despite their superficial adult similarities, they were quite different from *Alcippe* larvae and now he would have to go back and rethink *Alcippe* again as a consequence. He wrote to Hancock in frustration: 'This has utterly confounded my previous confusion how to rank Alcippe & it; for they present some most remarkable similarity, for instance they are both bisexual, with the males remarkably alike. & yet in what I must consider their fundamental organisation, & in their metamorphosis, they are so totally unlike that I cannot place them in the same order!'[3] Compare the adults and they seemed like doubles; compare the pupae and process of metamorphosis and they were as dissimilar as a lizard and a leopard.

Some of the analogies between these two deviants, though, *were* bizarre and wonderful. A week before, he had looked carefully at the cirri on the legs of the *Alcippe*, noting crenated ridges on them, like teeth. He had a hunch, a flash of inspiration, that these toothed ridges were not used for grasping but for tearing up food for digestion. So these were swimming legs with teeth – multi-functioning like a penknife. Then – wonder of wonders – only days later, searching inside Mr Arthrobalanus's oesophagus, the canal from mouth to stomach, he found, he wrote to Hancock: 'The most beautiful discs set with teeth, & brushes of hairs worked by muscles, certainly for triturating food; which strengthens my notion.'[4] So, despite the fact that these barnacles were not of the same species, nor of the same order – despite the fact that they came from different oceans – they had evolved similar bizarre devices for digestion: one had developed toothed legs and the other toothed stomach walls.

The minute Arthrobalanus males also had quite the largest genitalia he had ever seen in the barnacle world, he wrote in the manuscript, allowing himself a rarely used exclamation mark:

the prosbosciformed penis is wonderfully developed, so that in Cryptophialus, when fully extended, it must equal between eight and nine times the entire length of the animal! These males . . . consist of a mere bag, lined by a few muscles, enclosing an eye, and attached at the lower end by the pupal antennae, it has an orifice at its upper end, and within it there lies coiled up, like a great worm, the prosbosciformed penis . . . there is no mouth, no stomach, no thorax, no abdomen, and no appendages or limbs of any kind . . . I know of no other animal in the animal kingdom with such an amount of abortion.[5]

He had now identified barnacles with separate sexes in both of the major groups: both sessile (*Cryptophialus*) and stalked (*Ibla* and *Scalpellum*). It revealed spectacularly, he wrote, 'how gradually nature changes from one condition to the other – in this case from bisexuality to unisexuality'.[6] But these males were utterly and fully dispensable. They were minute, multiple, short-lived and had been reduced to only *three* segments of the usual seventeen barnacle body parts. This was enough for Arthrobalanus survival.

Though it shadowed the burrowing of *Alcippe*, and the separate sexes of *Ibla* and *Scalpellum*, Arthrobalanus seemed – extraordinarily – to belong to no one. He would have to have an order all of his own, a new and yet-to-be named order: *Abdominalis*, he decided, for, uniquely in the barnacle world, Arthrobalanus's legs were attached to its abdomen and not its thorax. He wrote up his final word portrait of *Cryptophialus minutus* at the end of March, summarizing the deviance and miracle of this creature in the passionless language of the systematist. There were no exclamation marks allowed here:

Cirripedia, having a flask-like carapace; body consisting of one cephalic, seven thoracic, and three abdominal segments; the latter bearing three pairs of cirri; the thoracic segments without limbs; mouth with the labrum greatly produced and capable of independent movements; oesophagus armed with teeth at its lower end; larva, firstly egg-like, without external limbs or an eye; lastly binocular, without thoracic legs, but with abdominal appendages.[7]

He was still uncertain, though. His instinct troubled him and he badly needed more time to think over the puzzle that Mr Arthrobalanus's abdomen presented. But he had no time. Lankester was pressing him hard. There was no option but to try to leave the subject as open as he dared, as he had done with the *Ibla* and *Scalpellum* descriptions. So he wrote a curious footnote, seeking to leave the door open as far as he could: 'I may add that I have several times tried to persuade myself, with no success, into the belief that I have somehow misunderstood the homologies of the thoracic segments and cirri of Trypetesa and Cryptophialus; for if this were so, the two genera could be brought into much closer relationship . . .'[8]

On 30 March he carefully placed the eleven prepared slides, with their black asphalt circles containing the dissected body parts of Mr Arthrobalanus, back in the narrow oak drawers of his cabinet, and pushed them shut. Case closed but – frustratingly – not to his satisfaction. Arthrobalanus was mapped but still only dimly understood in evolutionary terms. Time had run out.

He still had one final deviant to map: *Proteolepas*, which had no legs at all, neither attached to thorax nor abdomen. Like Arthrobalanus it burrowed, so was naked and resembled the larva of a maggot or a fly and like *Alcippe* had no stomach or anus. This too had to have its own separate order: *Apoda* – legless.

In 1835, only hours after he had picked up Mr Arthrobalanus's conch-shell home, whilst he dissected him in the cabin of the *Beagle*, the volcano out in the sea had begun to erupt spectacularly. He and FitzRoy had watched its angry lava pouring out into the night sky. Days later Darwin had felt the ground tremble under his feet on the shoreline of the South American forest and watched the sea surge up the beach, driven by submarine convulsions. Concepción had been devastated, the cathedral reduced to rubble.

Now, only three days after finishing the final dissections of Mr Arthrobalanus, he read newspaper reports of earthquakes in the west and south-west of England on the night of 1 April: rumbling noises, houses shaken as though by wind, windows rattling and bells ringing. These were but the aftershocks of a huge earthquake that was felt around the world that night, particularly in Canada and in the state of New York.[9]

Darwin was not a superstitious man; he saw no portents in earthquakes, no signs of divine unrest; but this had been a strange year, meteorologically: snowstorms, floods, fireballs, northern lights and now earthquakes.

In London, a few days later, at a meeting of the Geological Society, friends pointed out Thomas Huxley across the crowded room. Though still struggling to secure a university position,

25 Thomas Huxley

Huxley was making a career for himself, they said. The Toronto job had gone to another man, but he had already been given a Royal Medal for his work, and had delivered a Royal Institution lecture. From a distance Darwin watched him talking animatedly, gesticulating with his hands, black eyes flashing under thick black brows. Formally introduced, the two men opened a conversation about the sea squirt, a creature that, when disturbed, squirts sea water as it contracts. Huxley explained that he was cataloguing the British Museum's collection and had some ideas developing. John Chapman, he told Darwin, who enquired after his career, had offered him his own science review column in the heavyweight quarterly, the *Westminster Review*. This was power indeed, Darwin reflected. From the top of his column Huxley would be able to read and review all the latest science books and shape public reaction to them. If he came to be trusted by the readers of the *Westminster*, his word could make or break a theory or the reputation of a naturalist. Huxley's conversation was witty and sparkling, his language bold and colourful. He gossiped easily, was indiscreet. Darwin warmed to him. He had sea squirts back at Down House, he said quietly, rare ones that he would not be using. Would Huxley find them useful?

Immediately on his return to Down, exhausted from conversation and suffering from the painful flatulence that the London visits always brought on, Darwin wrote to Huxley to cement the beginnings of what he hoped might be a valuable friendship. He wrote to offer some twelve to fifteen sea squirts, describing in detail all their coils and colours and tails by way of bait. Tucked away on his shelves, he wrote, he had a copy of Johannes Müller's important work on echinoderms, a book that Huxley was unlikely to have seen and was bound to covet. So with a rare book and rare bottled specimens from the Falkland Islands promised but not yet sent, he slipped seamlessly into his own request. Rare sea squirts, it seemed, came with strings attached:

You spoke as if you had an intention to review my Cirripedia: it is very indelicate in me to say so, but it would give me great pleasure to see my work reviewed by any one so capable as you of praising anything which might deserve praise, and criticising the errors which no doubt it contains . . . it has been published a year, and no notice has been taken of it by any zoologist, except briefly by Dana. Upon my honour I never did such a thing before as suggest (not that I have exactly suggested this time) a review to any living being. . . . I have become a man of one idea, – cirripedes morning & night – [10]

Just days after receiving Darwin's letter, Huxley's mother died; but despite his grief, caused in part by his inability to find consolation in

Christianity,[11] he copied out and sent Darwin a copy of the mollusc paper he was working on, asking for comments and opinions and confirming that he would like to see the rare sea squirts. Bartering as well as conversation had begun and both men recognized the importance of the friendship. In the mollusc paper Huxley had declared his position on the species question – he was an archetype man, and his archetypes were absolute and unwavering – but at least he marked himself off from Richard Owen in this respect, by making it absolutely clear that these archetypes were real material structures like 'a diagram in a geometrical theorem',[12] not idealized Platonic blueprints. However – and this was a disappointment to Darwin – Huxley was clearly passionate about showing that nature had *absolute* archetypal boundaries. He was a fixity-of-species man.

This was difficult: how to get Huxley to see that species might be mutable without turning the man against him? How to keep the conversation going without becoming entrenched in battle positions? Darwin drew his most speculative prose about him like a cloak, as if he were thinking these things through, still not decided, an ignorant man groping for the truth:

I have read it all with much interest; but it wd be ridiculous in me to make any remarks on a subject on which I am so utterly ignorant . . . Several of your remarks have interested me; I am, however, surprised at what you say versus 'anamorphism': I shd have thought that the archetype in imagination was always in some degree embryonic, & therefore capable of & generally undergoing further development.[13]

Then in for the kill – seemingly light and inconsequential questions that Huxley would find difficult to answer.

Is it not an extraordinary fact, the great difference in position of the heart in different species of Cleodora? I am a believer that when any part usually constant differs considerably in different allied species; that it will be found in some degree variable, within the limits of the same species: – Thus, I shd expect that if great numbers of specimens of some of the species of Cleodora had been examined with this object in view, the position of the heart in some of the species, wd have been found variable. – Can you aid me with any analogous facts?[14]

This strategy came naturally to him. It was one he used with Hooker – rounds of carefully formulated questions which, when answered, began to work away at certainties. It would be some time before Darwin would succeed in converting Huxley to the species theory, but he had begun.

Now that the ground had thawed after the long snows and the floods had abated, Darwin turned from his barnacle burrowers to his own digging project. The pump from the now-abandoned douche had given him an idea for an experiment inspired by Justus von Liebig's *Familiar Letters on Chemistry*.[15] In London everyone was talking and writing about sanitary reform; without a proper sewer system such as the one Napoleon III was building in Paris, human waste would simply choke London, and the Thames was itself already no more than an open sewer. Everywhere people were devising schemes whereby clean water could be channelled into the city and the waste pumped out. Liebig was one of them. Passionate about circulating the chemicals of life and death, he advocated returning waste matter to the soil to rejuvenate it on a large scale. The followers of British sanitary reformer Edwin Chadwick were pressing for engineering schemes that would channel the waste straight into the fields, thereby providing free fertilizer, and setting up a perfect circulation between the putrefying and the living, the city and the country.[16] The huge machines and inventions of the Great Exhibition had made people feel that anything was possible with enough imagination, experimentation and financial investment.

While completing the dissection of his deviant burrowing barnacles *Alcippe*, Arthrobalanus and *Proteolepas*, Darwin wrote to his brother Erasmus:

I am very much obliged for the calculations about the Tanks. I am scheming a great water-work & heartily wish you were here to scheme: it is to make a very large tank; & then to be able from this to fill my three others, which are much smaller but as deep or deeper, from the shallow one. I want you to look when next at Athenaeum in Encyclopaedia, & see if you can find anything on the subject. The siphon I propose to be gutta Percha: it would have to be about 180 feet from top of the furthest tank to tank, not in a quite straight line . . . [17]

He had already built a liquid sewage tank ten years earlier at Down House,[18] and since then he had added sunken, covered brick-lined water tanks, so he knew how expensive these building schemes could be. He had some trading to do first, some cashing in of railway shares to release the money that would be needed to buy the bricks and the labour and that would pay for the design of the cables and pumps. If the summer was hot and dry, as it had been the previous year, the Down House gardeners would have a much greater supply of water to use on the vegetable and flower gardens and in the orchard.

Unable to work out the siphon system, he had written to the *Gardeners' Chronicle* at the end of April, explaining his scheme and asking for advice: 'Now can any one tell [me] whether a syphon made of Burgess and Keys' canvas hose, lined and coated with gutta percha, or of any other material, would practically answer? What bore should the siphon have, to convey in the course of 10 or 12 hours 3000 gallons of water?'[19]

The answers came not from the readers of the *Gardeners' Chronicle* but from Edward Cresy, to whom Darwin had written separately, the principal assistant clerk at the Metropolitan Board of Works and architect to the fire brigade, an expert on pumps and on gutta percha, a newly discovered natural rubber that had been used to line and insulate the great telegraph cable, laid three years before across the Atlantic seabed. At Cresy's suggestion Darwin travelled to the London offices of patent agents Burgess & Key to discuss the project on 7 May, combining his visit with Lord Rosse's Royal Society party, where he and Hooker talked late into the night about Hooker's travels in India.[20] Burgess & Key were not able to supply the piping that Darwin needed at a price he was prepared to pay, so despondently he settled on a much smaller project with a single tank, worked by the pump from the douche. It had taken up too much time already. The village builder, Issac Laslett, completed the work on 4 July.[21]

When George Sowerby came in mid-June to complete the illustrations for the 900-page third volume, Darwin worked 'like a slave'[22] alongside him, while outside Laslett completed the domed brickwork for the tank near the well. A promised holiday with

Emma and the children was now only a couple of weeks away. To make matters worse, a valuable collection of rare fossil barnacles had arrived from a chemist and fossil collector in Maastricht on 8 June. The new collection and the delays in completion of the garden tank[23] forced Darwin to postpone the family holiday. Elizabeth, who turned six on 8 July, and George, who turned eight on the following day, would have to celebrate their birthdays in Down House, not by the sea. He couldn't set off without at least seeing the tank complete, and the fossil barnacle collection was tantalizing; it was, Darwin wrote to Joseph Bosquet, 'a *magnificent* present . . . I truly hope that you have not robbed yourself'.[24] Finally, on 14 July, the fossil barnacle volume still frustratingly incomplete, Darwin and Emma made for Eastbourne, where Emma had secured lodgings for three weeks at 13 Sea Houses, in an elegant parade overlooking the sea.[25]

Darwin's mind was absorbed with digging, despite the completion of the tanks: now it was gold mines. Emma had been reading aloud from the volumes of Godfrey Charles Munday's new book *Our Antipodes . . . with a Glimpse of the Gold Fields*. The children loved to hear the gold-rush stories, particularly George, knowing that their father's friend and old servant Syms Covington lived out there in the great open spaces that Munday described. The discovery of gold in Western Australia had not led to the civil disorder that had so characterized the California gold rush; Munday described the gold mines as orderly and healthy places to make a fortune: 'there has scarcely been a case of serious sickness at either of the diggings. The scarcity of strong drinks, the plainest of food, physical activity combined with a healthy degree of mental excitement, seems to render drugs and doctors useless.'[26]

Munday's description of life in Sydney with its warm climate all year round, shark hunts, balls, picnics and oyster hunts, stood in sharp contrast to the Eastbourne in which they arrived in July 1853 as Darwin described it to Fox: 'Here we are in a state of profound idleness, which to me is a luxury; & we shd all, I believe, have been in a state of high enjoyment, had it not been for the detestable cold gales & much rain, which always gives much ennui to children, away from home.'[27]

The rains were torrential across the country that month, *The Times* reported. Many farmers who had lost livestock in the heavy spring snows, had been expecting an abundant hay harvest, which was now quite ruined. Darwin's additional tank, so carefully and expensively built to insure against drought, would not be drawn upon this year. In Eastbourne, tourists sheltered like the Darwins inside damp lodging houses and watched the rain fall on the bathing machines and donkeys and the colours change from green to grey out on the storm-mirrored sea. When Darwin asked Lenny how he liked Eastbourne, the boy, now three, replied, nodding towards the sea: 'I like that pond best but where will they put it to when we dig in the sand?'[28]

Darwin, free to watch his growing children adapt to a new environment, wondered at the variety of type and personality he and Emma had brought about as they promenaded with the children, governess and nurses gathered around them like a small tribe. William, released from the routines of school, enjoyed the freedoms of the family holiday and his own senior place in the family pecking order; Etty liked the seashore, collecting shells and showing an interest in identifying them with her father; Lizzie, now decidedly eccentric, talked and mumbled odd phrases to herself, keeping away from the other children, particularly the boys, whom she found tiresome; George, popular and good-natured, ran and chased and bothered the younger boys, played at soldiers and ignored his sisters. Lenny whined and complained a good deal about his bumps and cuts, so that George teased him about being a baby. Bony, they called him. He cried easily. Franky and Horace were like large, soft puppies, rolling and tumbling together, sand between their fat white toes.[29]

On long wet days, when Emma had finished reading about Australia and the gold rush, she transported the entire family to Africa for long hours by reading aloud from their cousin Francis Galton's new book which Darwin called *Tour in South Africa*. Galton, still in his early thirties, had sailed for Africa in 1850 in search of adventure and unknown territories and good shooting opportunities. He had found there 'shooting in abundance, and an opportunity to learn about an interesting race of negroes', the Damara tribe who lived in a land 'where no white man had ever penetrated'. The two-year journey, he wrote, had given him 'robust health' and had fostered 'habits of self-reliance in rude emergencies ... which are well worth possessing, though an English education hardly tends to promote them'.[30] Darwin dreamed of such adventures and experience for his own boys, fostering habits of self-reliance. It was the best book he had heard read for months, with its descriptions of the customs and rituals of the Damara tribe, as well as stories of mirages, missionary stations, night bivouacs in the desert, attacks by lions, shooting giraffes and watching game at night through a large pair of opera glasses – a manly tale told by a family member, blood stock. He wrote 'good' next to it in his reading notebook – so good, in fact, that he wrote to Galton from Eastbourne, care of the publishers, to congratulate him: 'What labours & dangers you have gone through: I can hardly fancy how you can have survived them, for you did not formally look very strong, but you must now be as tough as one of your own African waggons!'

In trying to tell Galton of his own life, however, he suddenly felt keenly aware of what such a life might look like from the outside to such a rugged and proven adventurer as his cousin:

'I live at a village called Down near Farnborough in Kent, & employ myself in Zoology; but the objects of my study are very small fry, & to a man accustomed to rhinoceroses & lions, would appear infinitely insignificant.'[31]

Natural modesty, but doubts too about the significance of the pile of papers that would soon, printed, make up the third of four volumes; he had made a mountain out of a barnacle. He found he couldn't even venture to tell Galton that the 'small fry' of his study were barnacles, the subject of his major work – his life's work, perhaps, if he never managed to finish the species book. Who would read such a book compared to those who would cherish Galton's story? And what would people say of the two of them? – Galton, the adventurer in uncharted Africa; Darwin, the barnacle man, the man of the footnote. Yet there was a frankness in Galton's style and the spirit of his book that Darwin felt he could learn from. His Preface made no grand claims: he offered simply the observations of a curious traveller into a world unknown to Europeans. He was reflective and enthusiastic; he admitted when he didn't know how things worked.

While in Eastbourne, Darwin and Emma read, to their horror, about the deaths of two little boys, aged eight and ten, who had been spending their summer at William Fox's rectory in Cheshire. These sons of family friends, having lost their mother, had been partially adopted by Fox and his wife, absorbed into their huge family. Ellen Fox had just given birth to their eleventh child and now they had scarlet fever in his household, the enemy within. Darwin and Emma knew what faced the distraught parents. They were under siege, waiting for the next strike, trying to explain the deaths to the small children, disinfecting everything, waiting. Darwin wrote to his cousin desperately from Eastbourne: 'do pray sometime tell me how far you have escaped'. But then Susan, Darwin's sister, forwarded an ominous letter from Fox reporting that their two-year-old daughter Louisa was seriously ill. Darwin wrote to Fox on 29 July:

'I am so sorry I sent off my former letter on indifferent subjects to you. – But the case has been incomparably worse than I had dreamed of. I did not know how completely the two Boys had been domesticated with you. – you have our deepest sympathy.'[32]

Louisa died that very day. When he heard the news some week or so later, when the family had returned to Down House, Darwin wrote immediately to his cousin, with the tight punctuation that he had always used when writing about Annie, as if the words choked him, made him stutter:

We too lost, as you may remember, not very long ago, a most dear child, of whom, I can hardly yet bear to think tranquilly; yet, as you must know from your own most painful experience, time softens & deadens, in a manner truly wonderful, one's feelings and regrets. At first it is indeed bitter. I can only hope that your health & that of poor Mrs Fox may be preserved; & that time may do its work softly, & bring you all together, once again as a happy family, which, as I can well believe, you so lately formed.[33]

Darwin had promised to take the children with Admiral Sulivan to see the military manoeuvres that had been going on all summer on Chobham Common in Surrey, which was conveniently only a handful of miles from Hermitage House, the home of Emma's brother and sister-in-law Jessie and Harry Wedgwood. Three days after writing to Fox, Darwin, Emma and family travelled to Surrey. The soldiers, camped here since June, for all their fine displays and mock battles, were preparing for war. Skirmishes between the Russians and the Turks were becoming critical along the Russian border, and war was likely to break out at any moment. Military leaders, ambassadors and politicians across Europe looked on anxiously, for any shift in the balance of power between these two mighty empires would have a global impact. For the British, increased Russian expansion in the Middle East threatened critical trade routes to India and to the Mediterranean. The balance of power had to be maintained. In July, whilst the Darwins were in Eastbourne in the rain, Russia had made its first move, marching into the principalities of Moldavia and Wallachia. War was now inevitable.

26 The Allied Camp on the plateau before Sebastopol, 1855

The sight of 10,000 soldiers storming enemy positions, practising tactics, building camps, bridges and defensive positions on this sandy heathland was, the papers claimed, the most dramatic spectacle of the season, comparable only to the Great Exhibition of two years earlier. It had to be seen. War fever and patriotism brought thousands of tourists that summer, including the Darwin and Sulivan families, with their combined clutch of thirteen children. Their first sight from the Wedgwood carriages was breathtaking: a two-mile stretch of levelled heath, studded with hundreds of white conical military tents, the whole panorama swept with wood smoke from camp fires. To Darwin's eyes, if to no one else's in the party, those conical tents must have looked for all the world like clusters of milky-white barnacles on a rocky shoreline at low tide – bizarre. Amongst the tents, thousands of human figures in red or white jackets moved about, cleaning and polishing equipment, cooking, tending the horses and digging trenches. Over the next three days the families stayed all day watching cavalry charges and thrilling to the sounds of metal on metal, rifle fire and the thundering of horses' hooves. On one day they found themselves in the path of a charging army, the 13th Light Dragoons, and 'had to run hard to get out of the way'. These same soldiers in dashing dark-blue uniform with gold braid, would charge with the Light Brigade at Balaclava in the Crimean War a year later, and many would die there.

George, eight, now completely besotted by all things military, spent his time showing Sulivan how he could spot the different regiments by the colour and cut of their uniforms: the Royal Horse Guards, the battalions of the 19th, 35th, 79th, 88th, and 97th Foot, 2nd Dragoons, 4th Light Dragoons and the 8th Hussars. From the vantage point of a hill Sulivan explained all the military movements with the greatest of approval as if it were a giant chess-board stretched out beneath them. The military strategist was in his element.[34]

Darwin's mind was also on strategy. He had a Preface to write and reviews to negotiate. How to draw the reader's mind from the chaos of barnacle detail to larger philosophical questions? And whether he *should*. He had originally intended to write a separate volume, in which he would consider all the philosophical questions the barnacles raised; but now, in this bellicose climate, he was not so sure he was ready to do so.[35] The still-anonymous author of one of the most speculative books of the century, *Vestiges of the Natural History of Creation*, had just brought out a tenth edition of his book. Darwin still thought it intriguing nonsense, but it had been in print for ten years and had brought a version of the development theory into the drawing rooms of Britain to be laughed at or taken seriously.

During those ten years, public opinion about development had shifted perceptibly, while he had been entirely caught up with his barnacles. The reviews of *Vestiges* were more positive now that the book had held its place in print for as long as this, and the reviewers seemed to want to engage with the larger questions of science, even if they remained critical of the inaccuracies in the elegantly written book, errors 'both in fact and philosophy'.[36] Huxley, too, though damning the book, noted in his review column that the questions 'thinking men' were prepared to ask were now bigger and bolder; but the climate of debate at the Geological Society and the Royal Society had become increasingly warlike on these big questions, Darwin noted, each time he went to London. War fever had infected all men of science, it seemed. Huxley spoke glibly in conversation about front lines and hostilities. Debates about development theories had come to sound like warfare, with intellectuals ranged on either side of a policed border, hurling bricks at each other. It must be possible, Darwin was sure, to put his ideas into words speculatively, tentatively, to engage in conversation and debate rather than intellectual and religious conflict.[37]

Now that London intellectuals were taking sides and journalists were championing the development hypothesis in the public eye, stirring up conflict, polarizing opinion, it seemed that no systematist in botany or zoology could easily avoid publishing new work without commenting one way or another. Prefaces and Introductions had become statements of position and conversations were being closed down this way, Darwin complained to the young naturalist, John Lubbock, working under his direction at Down. He was still waiting to see what Hooker would do with his first publication after their conversations at Down in the spring. The long-promised preview of Hooker's manuscript introduction to *Flora Novae-Zelandiae* arrived by post in September after a longer silence than usual between the two friends.[38] What would Hooker have to say about the species problem? The introductory essay, much longer than Darwin had expected, was brilliant, clever, well written. Hooker displayed his knowledge as a man of unquestionable authority, surveying recent developments in botany critically and judiciously.

In the opening pages Hooker wrote that he felt the need to declare his theoretical views on the origin, variation and dispersion of species, because all too often naturalists began research with a theory they wanted to prove that coloured their judgement. Whilst he felt humble about his ability to 'grapple with these great questions', Hooker wrote, he felt it necessary to do so. Then came six pages of suspense, while Hooker, having announced his intention to pin his colours to the mast, digressed to summarize the history of the botany of New Zealand. Darwin skipped anxiously through these pages, still uncertain how Hooker would stand on the 'great questions'. The first indications of a declaration-of-position critical statement began to unfold many pages later, as Hooker began to set out his stall:

> Although in the Flora I have proceeded on the assumption that species, however they originated or were created, have been handed down to us as such, and that all the individuals of a unisexual plant have proceeded from one individual, and all of a bisexual from a single pair, I wish it to be distinctly understood that I do not put this forward intending it to be interpreted into an avowal of the adoption of a fixed or unalterable opinion on my part.[39]

He had taken a position, he claimed, because a systematist 'should keep some such definite idea constantly before him, to give unity to his design'. For today, Hooker claimed, he was a fixity-of-species man. But tomorrow . . . ? He reserved the right to change his mind.

Darwin was not surprised at Hooker's decision to argue for fixity. After all, he had expected him to play safe with regard to species; he had no choice. Although Hooker might have wanted to allude to Darwin's natural selection ideas, he was bound to silence by his friend, and until Darwin actually *published* evidence about the mutability of species, Hooker would have to take the position that carried most evidence. Despite the arguments for fixity, however, Darwin found Hooker's discursive, open-minded Preface surprisingly powerful to read. His friend seemed to be thinking aloud, weighing up the evidence, speculating. He was opening up questions for discussion, not firing guns. The Preface might encourage others to do likewise:

dialogue, open discussion; not warfare. After all, it wasn't that Hooker absolutely believed in creationism – he wasn't fighting a Holy War – but rather that his friend's entire intellectual energy thus far in his botanical career had been dedicated to finding affinities, establishing stable patterns of kinship within the botanical world. He had a tidy mind, and within all the chaos of mountains of dried plant specimens in which he and Thomson worked at Kew, it perhaps seemed the only course open to them, to make sense of it all, to find patterns, mark out stable natural boundaries. Besides, his friend's career was still hanging in the balance, and he and Frances had started a family that spring.[40] Darwin had not gathered enough evidence as yet for him to risk that career for.

Darwin wrote to congratulate his friend on his 'admirable introduction', veiling his criticisms and defensiveness in the rhetoric of congratulation:

Many of your arguments appear to me very well put: & as far as my experience goes, the candid way in which you discuss the subject is unique. The whole will be very useful to me, whenever I undertake my volume; though parts take the wind very completely out of my sails, for I have for some time determined to give the arguments on *both* sides, (as far as I could) instead of arguing on the mutability side alone. [41]

He could not blame Hooker for arguing for the stability of species, but he might at least have alluded more to other hypothetical positions – he might have been more balanced perhaps. Hooker had made the evidence seem more clear-cut than it was; he skewed the picture in favour of fixity. Behind the scenes, in Darwin's study and in letters, hadn't Hooker admitted that on some days he couldn't decide whether the New Zealand specimens before him belonged to one species or twenty-eight? But then Hooker was approaching classification differently, impelled by the need to carve order out of the chaos of stamens and leaves and fibres. Darwin, on the other hand, was trying to do *two* things simultaneously. He was trying to classify species, as Hooker was, but he was also trying to understand *how* these species had evolved and diversified through eons of seabed and rockpool time.[42]

In the Preface, Hooker had, like Harvey before him, advised students of botany to undertake their investigations with no fixed idea to prove and without 'reference to any speculations which are too apt to lead the inquirer away from the rigorous investigation of details, from which alone truth can be elicited'.[43] Was this attack on men with a fixed idea an allusion to his friend Darwin? It was difficult to tell, Darwin reflected. He *had* started out on his barnacle research with a theory – that species had changed through time through a process of natural and sexual selection – and yet he had always been prepared to abandon it, if he had encountered incontrovertible evidence that disproved it. The species theory had complicated his barnacle findings, not simplified them. He felt he had to defend himself against Hooker's attack on men with fixed ideas, so he wrote carefully, trying to avoid sounding aggrieved:

> . . . in my own cirripedial work . . . I have not felt conscious that disbelieving in the *permanence* of species has made much difference one way or the other; in some few cases (if publishing avowedly on doctrine of non-permanence) I shd *not* have affixed names, & in some few cases shd have affixed names to remarkable varieties. Certainly I have felt it humiliating, discussing & doubting & examining over & over again, when in my own mind, the only doubt has been, whether the form varied *today or yesterday* (to put a fine point on it, as Snagsby would say). After describing a set of forms, as distinct species, tearing up my M.S., & making them one species; tearing that up & making them separate, & then making them one again (which has happened to me) I have gnashed my teeth, cursed species, & asked what sin I had committed to be so punished. But I must confess, that perhaps nearly the same thing wd have happened to me on any scheme of work.[44]

It was not that the theory stopped him seeing objectively, rather that no barnacle facts at any point contradicted his theory. Confirmation was everywhere; but always the classifier in him was in conflict with the speculator, he told Hooker. *One* looked to solve a problem one way, the other in another. Either way, headaches and sickness followed.

Whilst the words of Hooker's manuscript were not yet set in the stone of a printing room, Darwin knew that these finely chosen words were unlikely to change significantly. Hooker had stayed on the other side of the line – the fixity side. He had no choice while Darwin's theory remained a secret between them. Darwin was keen to absolve his friend tactfully of any feelings of responsibility, and he bade him farewell cheerfully, as if he were embarking on a voyage that would separate them for some time: 'Farewell, good luck to your work, – whether you make the species hold up their heads or hang them down, as long as you don't quite annihilate them or make them quite permanent; it will all be nuts to me; so farewell yours most truly, C. Darwin.'[45]

A week later, however, sending comments on the rest of Hooker's volume, he wrote more positively still, calling the essay 'perfect & elaborated . . . the most important discussion on the points in questions, ever published. I can say no more.' Yet it did make him feel gloomy, he admitted, to see that Hooker was about to put into print arguments for the permanence of species that at present he could not publically challenge and that amounted to hostile facts he would have to deal with later:

partly from feeling I could not answer some points which theoretically I shd have liked to have been different; partly from seeing *so far better done than I could* have done, discussions on some points which I had intended to have taken up . . . In a year or two's time, when I shall be at my species book (if I do not break down) I shall gnash my teeth & abuse you for having put so many hostile facts so confoundedly well.[46]

He consoled himself, he told Hooker, by reading Elizabeth Gaskell's new novel *Ruth*, which he asked Hooker to tell his wife, was 'quite charming'. 'I am becoming an abandoned novel-reader,' he added, conscious that Frances was keen to convert them all to novel-reading.[47] He abandoned his plans to write an accompanying volume in which he would tease out all the philosophical conclusions to which the barnacles had carried him.[48] He would, he decided, let the barnacles speak for themselves. When he did take a position, it would not be here in the pages of the barnacle research. It would colour readers' reactions to it, make it seem that he was a man *so* fixed in his tracks that he *would* prove one idea only. He would stay silent on the bigger issues, perhaps occasionally infer a philosophical conclusion but never state one categorically. The battle – if it proved to be a battle – would not be fought on barnacle ground.

A month later, in November, Hooker wrote effusively from London to congratulate Darwin on winning the Royal Medal for his work on the stalked barnacles, admitting playfully, however: 'I neither proposed you, nor seconded you; nor voted for you.' The prize, a gold medal of some considerable weight, would be presented on 30 November. The 'warmth, friendship & kindness' of Hooker's enthusiasm made Darwin 'glow with pleasure till my very heart throbbed', a pleasure that was worth more, he wrote to Hooker, 'than all the medals that ever were or would be coined'.[49]

In early December he carried the third barnacle volume on sessile barnacles, containing his sketchy conclusions about Mr Arthrobalanus, to London – the longest and heaviest manuscript of them all, amounting to almost a thousand pages; not a manuscript to trust to a servant to deliver, even one as trustworthy as Parslow, nor a manuscript to lose sight of until it was handed over to Edwin Lankester personally.

The proofs began to arrive at the beginning of February 1854 and at the same time the Palaeontographical Society confirmed that they did want him to complete a separate volume on the fossil sessile barnacles for their series, despite Darwin's concerns that there was really very little to say about the small number of fossil specimens that were of a decent enough condition to study.[50] As he completed the proofs and tried to describe the last fossil specimens, he plunged into Hooker's *Himalayan Journals*, delighted to find that Hooker had dedicated them secretly to him. Since he had read and reread Darwin's wonderful *Beagle* travels, Hooker wrote to him, it had been his life's ambition to complete such a book of exploration himself and 'I am now happy to go on jog-trot at Botany till the end of my days'.[51]

Britain declared war on Russia in March. Sulivan enthusiastically reported for duty and sailed for the Baltic in command of his own ship, the *Lightning*, on the twenty-fifth. Most reports expected the military action to be over by the following Christmas. Darwin was full of anxieties about the consequences of war with Russia, fearful about completing the barnacle volumes and moving on to the inevitably controversial species book. Would his species theory explode like a bomb, leaving a mark on the world for ever? Or explode in apparent emptiness, ejecting a million spores into the air? He wrote to Hooker: 'How awfully flat I shall feel, if when I get my notes together on species &c &c, the whole thing explodes like an empty puff-ball.'[52] Within weeks, despite his fear that he had lost Hooker to the other side, the two men were in discussion about the species theory again, Hooker asking more and more questions of Darwin about what he understood by high and low, and teasing Darwin about his 'elastic theory of creations & perfections & imperfections'.[53]

By July the proofed third volume was whirring its hundreds of pages through the presses of Charles and James Adlard, Printers, in Bartholomew Close, London, and the short fossil barnacle volume was nearing completion. In September 1854 Darwin could see an end to what had come to feel like damnation. He wrote to Hooker promising him a complimentary copy of the third volume:

I am very glad you wish to have my Barnacle Book, for I would rather send it to you than to any half-dozen-others, if you cared to have it. Our old friend Arthrobalanus is now christened Cryptophialus. Under the Order to which it belongs, I discuss the (as it appears to me) very curious case of its affinities; I was most uncomfortably puzzled how to class it & am far from sure that I decided correctly.

I have been frittering away my time for the last several weeks in a wearisome manner, partly idleness, & odds & ends, & sending ten-thousand Barnacles out of the house all over the world. – But I shall now in a day or two begin to look over my old notes on species. What a deal I shall have to discuss with you: I shall have to look sharp that I do not 'progress' into one of the greatest bores in life to the few like you with lots of knowledge.[54]

12

The Universe in a Barnacle Shell

*

To see a world in a grain of sand
And heaven in a wild flower
Hold infinity in the palm of your hand
And eternity in an hour.

William Blake, 'Auguries of Innocence'

Late October 1854; early evening. Darwin's study looks bigger now that he has banished his barnacles either to the British Museum or back to their collectors across the world. Many, shrouded in pill boxes, are still making their way through the British or European postal systems in the luggage compartments of railway carriages or sailing across night seas in the holds of ships.

Darwin, in a brightly coloured dressing gown, sits at a desk on which four books are arranged side by side; their polished leather surfaces catch the soft light from the oil lamp.[1] There are two thick volumes and two slim ones. The two thick volumes have the gold letters 'RS' embossed on the front cover, the mark of the Ray Society. Darwin runs his finger around the intricate design. There is also a pile of loose, thick milky-white papers on his desk – the 1844 version of the species essay, copied out for him in September 1844 in a delicate, small hand by Mr Fletcher from a nearby village.[2] It has been locked away for ten years. The ink is blue – cobalt blue – and Darwin has already begun to annotate the script. He has also prepared another list of questions for Hooker.

Joseph and Frances Hooker are here in Down House staying for a few days with Charles and Mary Lyell.[3] John Lubbock has been invited, too, and Darwin listens out for the familiar sound of his pony's hooves on the gravel outside. The house guests are all upstairs, dressing for dinner, but Darwin has slipped away to the study to look again at the full set of four barnacle volumes, which he plans to show Hooker and Lyell after dinner. On his shelf they take up more space than the combined width of his other four published books – the books on coral reefs, the *Beagle* voyage, volcanic islands and South America. Several weighty inches of shelf space. Eight years of his own close work; millions of years of nature's seabed metamorphoses mapped. These wretched barnacles had confounded him, driven him to distraction so that, at times, he had come to hate them; but now, looking through the plates interleaved into the Ray Society volume, he can see again how beautiful they are. Right now around the world, in darkened rock pools, in the coal blackness of the seabed, on the coppered bottoms of sailing ships and on the fleshy sides of whales, barnacles feed in millions upon millions on plankton, pulsing their jointed, delicate, feathery legs, like an aquatic orchestra.

What has he achieved, he wonders now, in bringing light to this small corner of the living world? He has produced a definitive monograph, he knows – probably one that will continue to be used by cirripedologists long after his death. The books are unique, the sum of all barnacle knowledge; they have pushed out the frontiers of the known world. But what have they done for *him*? Who is *he* now that he has finished the barnacle labours? A changed man without doubt. He thinks differently, reasons differently. Classifying the barnacles has given him new skills, language and understanding; they have sharpened his mind and his comprehension of theoretical principles in zoology, embryology and homology in particular; they have forced him to confront and solve problems of nomenclature; they have made him a skilled dissector.

More than anything, they had enabled him to ruminate over his species theory and strengthen his grasp of natural and sexual selection. The search for means of survival and reproduction had taken these barnacles *every which way*, since their earliest ancestral forms: some had developed thicker shells for defence, others faster unfurling of cirri for feeding; here a species was hermaphrodite, there it had begun to develop separate sexes. The variation of body structure across the group as a whole was extraordinary. Barnacles had shown him that it was almost impossible to mark a line where a variation with species stopped and where distinct species began – nature produced no such lines of absolute demarcation. When he began, he had been tormented by these doubts about demarcation. Now he had come to terms with this wateriness out there; he had come to understand that variations between one form and another 'blend into each other in an insensible series'. 'Trace gradation between associated & non associated animals. – & the story will be complete,' he had written in 1837. Now he understood that a small variation in a valve or leg was the first step that marked an incipient species: a small variation of that kind would lead to a well-marked variation, thus to a subspecies and eventually to a species.[4] Blending, seeping, mutating, nature was – and always had been – incontrovertibly on the move but imperceptible to the naked eye, like Lyell's hour hand.[5]

The world of zoology had changed, too, in these eight long years. It had been a period of intense zoological enquiry for scores of naturalists, and many had been working away like him on sea creatures, examining the dark world of rock pools and the seabed, creatures made visible on glass slides and the lenses of microscopes. Darwin's barnacle volumes were only a small part of a new push into the unknown undersea world. Most of these writers were men whom Darwin either knew personally or with whom he had corresponded over the previous years – men such as Dana, Huxley, Hancock, Bate, Forbes and Bosquet, meticulous observers like himself. Almost all of them were listed in the footnotes of his barnacle volumes – the hidden labourers – now they were underground, down there in the subterranean depths of his book.

What had been the result of their investigations into the watery unknown? Back in 1837 the Revd Jenyns had insisted that each naturalist should specialize on one area of zoology, then the combined collection of data would begin to tell its own story. Now early twenty years had passed since Jenyns had urged young naturalists not to *speculate* on nature's ways without first minutely describing them. Speculations about the origin of life or the relationship between the 'higher' and the 'lower' organisms must, he argued, be built only on the bedrock of extensive empirical observation. A speculator must be – or have been – an accurate systematist. Hooker had reaffirmed this principle even in 1845 when he had written to Darwin that 'no one has the right to examine the question of species who has not minutely described many.' Darwin felt the truth of this observation acutely in 1846. The barnacles would win him that right.

When he carefully pieced together the Preface to *On the Origin of Species by Natural Selection* in 1859, it was precisely this right that he claimed when he declared that these years of hard systematic observation and analysis proved that he had not been 'hasty' in coming to his decision: 'After five years' work I allowed myself to speculate on the subject, and drew up some short notes; these I enlarged in 1844 into a sketch of the conclusions, which then seemed to me probable: from that period to the present day I have steadily pursued the same object. I hope that I may be excused for entering on these personal details, as I give them to show that I have not been hasty in coming to a decision.'

Eight years winning a reputation as a brilliant and dogged systematist had brought Darwin other advantages. It had bought him time. In the barnacle years, the weight of systematic work undertaken in botany, zoology and comparative anatomy had brought the question of the permanence of species to centre stage. For, during these years, like Hooker with his New Zealand plants and Darwin with his barnacles, naturalists trying to systematise nature had been confounded by decisions about where variations within a species or subspecies ended and new species or subspecies began. In the Preface to the third edition of *The Origin of Species* Darwin listed the thirty-four naturalists who had published on the impermanence of species since 1800. *Ten* of these published their claims between 1846 and 1854. The idea of the fixity of species came under considerable pressure, then, during the barnacle years. Darwin's species theory would be read differently as a consequence of such increasing pressure.

Darwin also made an intricate web of correspondence through the barnacle work, a web established by gentlemanly good will, specimen exchange and philosophical and theoretical debate. In the thousands of letters he wrote during the barnacle years, he had written references, requested specimens, congratulated naturalists on their recently-published work, corrected errors in manuscripts, asked questions, answered questions, enquired after family and friends, cajoled, bantered, bartered, flattered and criticised. As a man of independent means with considerable charm and humility, he had made no enemies and kept himself remarkably unentangled in the politics of academic institutions. He had been overwhelmed, he wrote in the Preface to the book on the stalked barnacles, by the generosity of his fellow naturalists: 'if a person wants to ascertain how much true kindness exists amongst the disciples of Natural History, he should undertake, as I have done, a monograph on some tribe of animals, and let his wish for assistance be known.' And of course Darwin, contributing a specimen here, a reference or contact there, was himself only a skein in the knowledge webs of other naturalists working on similar projects. The elaborate global epistolary web to which he was joined would help to shape the reception of the species book.

Darwin had decided *not* to add an additional volume, or a long introduction, as Hooker had done, explaining what all of this weight of barnacle anatomical detail meant philosophically. He had determined to keep his barnacles publicly unframed by his species theory. He would keep his systematic barnacle work largely unspeculative, his speculation for a separate performance. Although the development hypothesis was now much more widely known and discussed, it was still inextricably linked to the brilliant and frustrating book published in 1844 as *Vestiges*,[6] which was now in its tenth edition and had sold hundreds of thousands of copies. *Vestiges* had helped to popularize questions about the laws of nature and the origins of life, but at the same time, because it involved grotesque speculation and was implicitly materialistic, it made the development hypothesis easy to demolish and ridicule. Its author just hadn't done the painful watching and recording and dissecting that was necessary. He hadn't earned the authority to speculate. Thomas Henry Huxley had published a savage review of the tenth edition of *Vestiges* in the pages of *The British and Foreign Medico-Chirurgical Review*. It was a review that made Darwin shudder. He could remember the opening sentences word for word: 'In the mind of any one at all practically acquainted with science, the appearance of a new edition of the "Vestiges" at the present day, has much the effect that the inconvenient pertinacity of Banquo had upon Macbeth. "Time was, that when the brains were out, the man would die."' [7]

Darwin, acutely aware that Huxley had promised to review his own barnacle works, had already written to him about the *Vestiges* review, which, he wrote, was: 'incomparably the best review I have read on the Vestiges; but I cannot think but that you are rather hard on the poor author. I must think that such a book, if it does no other good, spreads the taste for natural science. – But I am perhaps no fair judge for I am almost as unorthodox about species as the Vestiges itself, though I hope not quite so unphilosophical'. [8]

Darwin did not need to worry about Huxley's opprobrium, however. Whatever clues the barnacle volumes might have carried about Darwin's big idea, as yet undeclared, Huxley found them impressive and inoffensive. The books confirmed Darwin's rank as a systematist of international standing, Huxley declared:

Mr Darwin's present work shows him to be as able an observer of nature on the small as on the large scale. It deals with the anatomy and metamorphoses of certain crustaceans, those well-known barnacles, in all their varieties, from those which infest the bottoms of our ships, to those which lodge in the skins of Leviathan himself. Blind, fixed, and helpless as they seem to us, these animals in their young state, are active, sharp-sighted little creatures, somewhat like our water-fleas, with long leg-like antennae, provided with cups at their extremities. A time comes, however, when they know that they have to settle down in life; they adhere to some fixed or floating body by their sucking cups; then a long hump – we can call it nothing else – somewhat like that with which Mr Punch is provided, only ten times as long, grows out of their backs. From the end of the hump a sticky cement is poured out, which glues them firmly to their support; the function of the sucking-arms thus cease, but, as Mr Darwin has made out, they remain during life the witnesses of a different state of existence. [9]

Huxley, writing for the educated readers of the *Westminster Review*, had compared Darwin's books to the most important zoological studies of Europe, and at the same time had exploited the comic potential of the barnacle by turning it into a Mr Punch, a theatrical zoological sideshow. He anthropomorphized the barnacle as a man settling down, gluing himself to a rock. Although he wouldn't admit it in the review, Huxley was in the process of metamorphosing himself, trying his utmost to shift from a free-swimming form to a cemented one; he couldn't marry his fiancée Nettie until he had secured the job that would make married life possible. In 1854 permanent employment was within his grasp and his metamorphosis would be complete.

Sea creatures were worth watching; they were both entertaining and philosophical and, now that the aquarium had been invented, it was possible to do so. The *Westminster Review*, committed to progressive ideas, had been interested in the philosophy of sea creatures since the relaunch of the journal in 1852 under the editorship of John Chapman and Mary Ann Evans (later George Eliot). Evans had commissioned Professor Edward Forbes to write a splendid piece called 'Shell-Fish: Their Ways and Works'. In the journal it sat sandwiched between an article on representative political reform and one on the relationship between employers and the employed. This was one of Forbes's most lyrical publications yet – 'there is a philosophy in oyster-shells undreamed of by the mere conchologist!' he wrote; 'a noble and wondrous philosophy revealing to us glimpses of the workings of creative power among the dim and distant abysses of the incalculable past ... unfolding for us the pages of the volume in which the history of our planet, its convulsions and tranquilities, its revolutions and gradualities, are inscribed in unmistakable characters.'[10]

Like the oyster, the story of the barnacle was a story of 'convulsions and tranquilities ... revolutions and gradualities'; for Darwin, it was a tale impossible to tell or explain without recourse to the development hypothesis. This tiny creature had both a life cycle and an adaptation since prehistory that were as epic and spectacular as the story told in the pages of the *Vestiges of the Natural History of Creation*; it also had a life history that bizarrely shadowed patterns of human life, shaped as it was by the same natural laws of survival, development and reproduction.

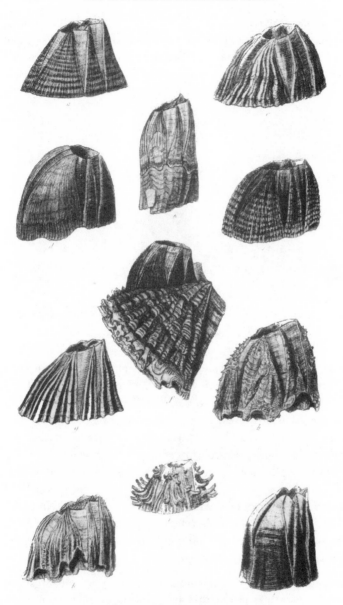

27 The Outside of an Acorn Barnacle

So if sea creatures had, by 1854 when Darwin finished his books, become engorged with philosophical questions, they had also begun to capture the imagination of the British public, particularly now that the spread of the railways had made seaside excursions and seaside

28 'Common Objects at the Seaside'

holidays so fashionable. Gosse had followed up *A Naturalist's Sojourn in Jamaica* with a much more popular book two years later called *A Naturalist's Rambles on the Devonshire Coast*.[11] Still committed to finding ways to study live marine creatures in conditions close to their natural habitats, he had by 1853 worked out a chemical formula for reproducing artificial seawater that supported marine life. That year the keepers of the Zoological Gardens, working closely with Gosse, had opened an aquarium, which had become one of the spectacles of London. These tanks, filled with seawater carried from the North Sea to London by the Great Eastern Railway Company, were theatres of glass in which sea creatures performed epic natural dramas, as John Timbs described in *Curiosities of London*:

The Aquatic Vivarium, built of iron and glass, in 1853, in the south garden, consists of glass tanks, in which fish spawn, zoophytes produce young, and algae luxuriate; crustacea and mollusca live successfully, and ascidian poplypes are illustrated, together with sea anemones, jelly fishes and star fishes, rare shell-fishes &c.: a new world of animal life is here seen as in the depths of the ocean, with masses of rock, sand, gravel, corallines, sea-weed and sea-water; the animals are in a state of natural restlessness, now quiescent, now eating and being eaten.[12]

29 'Valuable Additions to the Aquarium'

As Darwin rewrote and expanded the species theory essay into the book he would publish in 1859, called *On the Origin of Species by Natural Selection*, aquarium mania began to sweep the country. By this point Darwin was compelled by new questions and new experiments, and pumping Hooker and his newer correspondents – seed growers, horse, cattle and pigeon breeders and nurserymen – with endless questions about plant and animal distribution. In March 1855 he was concerned to understand the *means* by which botanical species spread from continent to continent: wind, seawater, rafts; in the stomachs of birds, or on the feet of birds. 'Really these questions are like Cerberus & his heads,' Hooker wrote in response to another set of questions posted to him by Darwin, 'the more arguments one disposes of the more rise up in grim array.'[13]

When Darwin met Gosse for the first time on 2 March at the Linnaean Society meeting in London, his mind was full of questions and problems about the distribution of species between continents.

There was no way to settle some of these questions without extensive and carefully described experiments with living species. His meeting with Gosse was timely, for the naturalist told him of his own experiments with seawater tanks in his home in St Mary Church, Devon, the tanks now successfully moved to London. He'd invented a formula for artificial seawater, he said, that made these experiments much easier because now he wasn't dependent on seawater supplies being shipped up the Thames. The recipe he had finally fixed on was a quarter-ounce of Epsom Salts to three and a quarter ounces of common table salt with 200 grains of chloride of magnesium and 40 grains of chloride of potassium, all mixed into four quarts of water. He was using water from London's New River suppliers, he said, one of the purest of the London water supplies. If Darwin wanted to undertake experiments with seawater, he should write to Mr Bolton of Holborn, a chemist who had agreed to market the salt mixture in special packets.

Gosse had written all these instructions out clearly for his readers in a new book he'd just published, called *A Handbook to the Marine Aquarium*, a sequel to his earlier book *The Aquarium: an Unveiling of the Wonders of the Deep Sea*. It took Darwin several days to see what Gosse's invention had made possible for him. Several weeks later he was still excited. He wrote to Henslow: 'I saw Mr Gosse the other night & he told me that he had now the same several sea-animals & algae living & breeding for 13 months in the *same* artificially made sea water! Does not this tempt you? It almost tempts me to set up a marine vivarium.'[14]

After all, Darwin had only studied *dead* barnacles. How much more there might be to learn with live ones.

It was Hooker who was most in Darwin's mind as he thought through his seawater experiments, for Hooker had always said that seeds of freshwater species couldn't travel in seawater because the water would kill them. They wouldn't germinate after a sea voyage. With seawater now so easily available through Gosse's artificial formula, Darwin realized that he could prove that seeds *could* be carried by seawater from island to island and continent to continent, by a series of experiments that no one had yet undertaken. The experiments would be extensive, he told Emma. He would need to enlist the labour of the entire household, including the children.

At Darwin's request, Henslow posted a parcel of waterweed from the ponds around Hitcham. First he had to establish a way of keeping the freshwater organisms, to be used in his experiments, alive. Emma put together a glass tank with gravel and water from a local pond, and when the waterweed arrived she slipped its roots into the gravel. At Gosse's advice they waited several days for the plant to establish itself and aerate the water.[15] Meanwhile, Emma, already skilled as a chemist in her role as supplier of remedies for the villagers of Down, made up Gosse's recipe for artificial seawater when the chemicals arrived in the post from Mr Bolton, 146 Holborn Bars.[16]

Darwin and Etty dropped and sprinkled cabbage, radish, spinach, oat, barley, borage, beet and canary seeds into small bottles, tanks or saucers filled with the artificial seawater. Darwin made lists of all the seeds and labelled the saucers carefully, giving George and Etty some limited responsibility for daily checking. The servants carried the saucers to the designated locations on Darwin's list – window sills in the schoolroom, drawing room, study, kitchen – and they measured the temperature in each room, carefully recording it in the notebooks Darwin gave them, so that he could check the effect of different temperatures on germination. The children, excited about competing against Hooker, eagerly watched for the results, keeping notes, watching the saucers in the schoolroom and in the drawing room, checking several times a day for the faint green flush which would confirm germination.[17] Despite the foul smell of the water, almost all the seeds germinated after between seven and fourteen days, to the children's triumphant delight.[18]

At the same time Darwin began a series of related experiments to see if freshwater animal species and their eggs would survive in seawater, writing to John Lubbock to request some freshwater molluscs, which he introduced into the tanks. By May he had decided to try lizard eggs and land snail eggs in seawater, but lizards' eggs were particularly difficult to procure, so he wrote to the *Gardeners'*

Chronicle for assistance: 'If any of your readers could obtain for me some eggs of the Lacerta agilis, I would be greatly obliged. Lizards are most widely distributed, and I want to ascertain whether the eggs will float in sea-water, and, if so, whether they will retain their vitality. A reward of five shillings . . . offered to schoolboys, would perhaps get these eggs in the proper districts collected.'[19]

In June he was still watching the seeds and giving Hooker regular bulletins about the results. He wrote to tell Miles Berkeley, a clergyman and botanist, who had undertaken similar experiments with seeds, that he had one set of seeds that took fifty-six days to germinate. This was excellent news – seeds could travel a considerable distance in fifty-six days and still be washed up on a shoreline with a chance of germinating; but Hooker was still not conceding ground, Darwin complained to a friend – he wanted yet more evidence: 'Hooker seems much interested in these experiments; but they seem to have had very little influence, or no influence, in making him think that plants thus get distributed, which I am rather surprised at; & I shd like sometime very much to hear your opinion on this head.'[20]

Thousands of seeds would *not* germinate, but amongst them and against all the odds a few rare seeds would burst into green shoots in the most hostile of conditions: seeds with swimming legs, seeds with the ability to withstand the effects of seawater, seeds that cemented themselves to rocks. Nature had devised extraordinary means of survival amongst all her carnage. Seeds and seawater had been at the heart of his experiments with Robert Grant on the seashore of Leith in the 1820s; seeds and seawater were at the heart of his experiments in Down thirty years later.

Darwin was gathering all these infinitely small facts as part of the evidence needed to support his species theory, for now that he had demonstrated his ability to observe and map nature on the very smallest of scales, he had returned to the most ambitious and epic projects of all, as he described it to his cousin William Fox:

I forget whether I ever told you what the object of my present work is, – it is to view all facts that I can master (eheu, eheu, how ignorant I find I am) in Nat. History, (as on geograph. Distribution, palaeontology, classification Hybridism, domestic animals & plants &c &c &c) to see how far they favour or are opposed to the notion that wild species are mutable or immutable: I mean with my utmost power to give all arguments & facts on both sides. I have a *number* of people helping me in every way, & giving me most valuable assistance; but I often doubt whether the subject will not quite overpower me. [21]

Gosse, without realizing it, had been enlisted on to the international team who were helping Darwin with the series of experiments that would result in the publication, in 1859, of *Origin of Species by Natural Selection*, a book that would trouble him profoundly and overpower him, for whilst Gosse the scientist would thrill at Darwin's arguments and evidence, Gosse the fundamentalist member of the Plymouth Brethen would quail at its implications. [22]

The barnacle years had been no cul-de-sac in the development of Darwin's ideas. Mr Arthrobalanus and his tribe helped him fine-tune the way he used homology and embryology to think about species' origins and relations, they had provided the foundation on which he had continued to build his credibility, reputation and authority as a systematist, and they had been the means by which he had established a network of correspondents that would hold together and bolster that authority.

The relentless task of putting barnacles into words also crucially sharpened Darwin's writing abilities both in terms of the need for hard, uncompromising accuracy in describing the curve of a valve, the texture of a shell, the colour of an oesophagus, or the striated ridges of the cirri, and in terms of the need for rhetorical hesitancy when the evidence for a hypothesis was as thin as it was with the *Ibla* complemental males for instance: 'it might reasonably be assumed that', 'there is some evidence to suggest that', 'it could be assumed that'. Such rhetorical dexterity, the ability to craft sequences of clauses that moved relentlessly but cautiously from evidence drawn from the commonplace creatures such as pigeons or barnacles to extraordinary hypotheses, underpinned by modest, tactful common sense, would be crucial to the success of the *On the Origin of Species by Natural Selection*. Through such sentences as these, Darwin could appear at once hesitant, judicious and abso-lutely certain as he was in the opening pages of *The Origin*:

Although much remains obscure, and will long remain obscure, I can entertain no doubt, after the most deliberate study and dispassionate judgement of which I am capable, that the view which most naturalists entertain, and which I formerly entertained – namely, that each species has been independently created – is erroneous.

In 1851, a young American journalist, Herman Melville, had published a book in England entitled simply, *The White Whale*, already published in America as *Moby Dick*. Its English title caused confusion for some readers who assumed it was a natural history of the whale. In it he wrote about his struggle to put the white whale into words. Though Melville was writing about a whale, he might have been describing Darwin's encounter with a creature too small to see with the naked eye but which had taken a brilliant man eight years to comprehend:

One often hears of writers that rise and swell with their subject, though it may seem but an ordinary one. How, then, with me, writing of this Leviathan? Unconsciously my chirography expands into placard capitals. Give me a condor's quill! Give me Vesuvius' crater for an inkstand! Friends, hold my arms! For in the mere act of penning my thoughts of this Leviathan, they weary me, and make me faint with their outreaching comprehensiveness of sweep, as if to include the whole circle of the sciences, and all the generations of whales and men, and mastodons, past, present, and to come, with all the revolving panoramas of empire on earth, and throughout the whole universe, not excluding its suburbs. Such, and so magnifying, is the virtue of a large and liberal theme! We expand to its bulk. To produce a mighty book, you must choose a mighty theme.

Now that the barnacles were classified, mapped, crossed every which way, Darwin could turn his mind to the species theory – his mighty book – waiting in the wings. He could no longer prevaricate, bury himself in his mountain of minute facts. 'I am like Croesus overwhelmed with my riches in facts', he wrote to his cousin, '& I mean to make my book as perfect as ever I can.' Now those facts needed to be marshalled, put to use, assembled to support a revolutionary theory that would be at once careful, reasoned and tactful. Now, pen in hand, he would begin to make his transition from one Herculean task to another.

The Asphalt Curtain

*

Without our improved microscopes, and while the sciences of comparative anatomy and chemistry were yet infantile, it was difficult to believe what was the truth; and for this simple reason that, as usual, the truth, when discovered, turned out far more startling and prodigious than the dreams which men had substituted for it, more strange than Ovid's old story that the coral was soft under the sea . . .

Charles Kingsley, *Glaucus; or the Wonders of the Shore* (1855)

A seventy-foot whale skeleton is suspended in mid-air above the entrance of the Cambridge Zoology Museum, hanging from wires against its modern concrete façade like a giant puppet. The vaults under the public museum contain important Darwin *Beagle* specimens, including the octopus that Darwin played with in the South American rock pool and then preserved in a glass jar. Dr Friday, lecturer in Zoology and museum curator, agreed to hunt it out for me to see. When he met me in the entrance hall, I asked him about the whale. The skeleton had nearly killed him, he said. In the 1960s, when the new building had been completed, he and fellow curators had devised a system for reassembling it in mid-air. Each bone had been labelled and special bolts, pulleys and wires designed and assembled. He had volunteered to work up on the scaffolding making sure that each bone part was properly bolted together in the right order. It was only when the team had almost completed the skeleton that he had realized he had inadvertently been bolted into the ribcage, his head between two of its ribs, now suspended in mid-air. He had laughed, then slipped, and it had only been by reaching out for a hanging bone that he had prevented himself from being garrotted between whale ribs.

We made our way down through the public museum – all white-washed walls and glass, weaving our way through the gleam of the ivory-coloured skeletons that pressed upon us on every side – giraffes, geckos, hippos and antelopes. In glass cases bleached bird skeletons sat on nests with their young. Lizards, salamanders and monkeys would have peered back at us if they had had eyes to do so.

We crossed the threshold marking the boundary between the public and the private museums, passing through a door that appeared to open out of nowhere, down on to a narrow, steep flight of steps; white breeze blocks, brightly lit, concrete floors. Occasionally another zoologist or researcher in a white coat would brush past us and disappear into another locked room or down another corridor. On the tops of green metal storage cupboards, the skulls of antelopes and tigers flanked by ancient wooden trunks marked 'mammal skins' and '*Marsupialia*' stared down at us blankly. Now we were in single file and my guide was concentrating on finding the right keys with which to open the series of locked doors that lay ahead of us. We were, he said, going to 'The Spirit Store'. It sounded like a phrase from a Gothic Romance or from a novel by Gabriel Garcia Marquez. For a moment I imagined a room crowded with ghosts of animals and people, flickering; but it was where they kept bottled specimens, he explained, preserved in spirits of wine. The air was thick with alcohol, evaporating constantly from the bottles. Despite the air-conditioning system, we would be breathing in alcohol evaporated from the ancient bodies of snakes and snails and lizards. He had the store number for Darwin's South American octopus, he said, and he should be able to find it quickly. Then, when I had finished with the octopus, he had something unusual to show me.

In the Spirit Store we had to raise our voices over the whirr of the air-conditioning system. I could smell the alcohol in the air. I could taste it. Green metal shelves made long corridors down the room, each shelf crowded with bottles of different sizes and shapes – modern Kilner storage jars, or Victorian sweet jars, or jam jars, or perfume bottles – containing pickled specimens of every imaginable creature, suspended in pale amber fluid: from sea slugs, armadillos and lizards to moles and marmosets. Suckers, coils, eyes, feet: suspended, embryonic and immortalized. Some specimens were from the *Beagle*; others had been donated from other nineteenth-century collections. The glimpse of a dismembered snake's head, jaw open fiercely, seen out of the corner of my eye, made the hairs stand up on the back of my neck. The lights flickered intermittently. My guide had disappeared in search of the bottled octopus.

30 The Spirit Store

Darwin's handwriting was just visible on the label of the jar placed before me at eye level. It was handwriting I knew well but which I had not expected to see here in this strangest of places. The handwriting conjured the man himself, cooped up and nauseous in the cabin of the *Beagle* in 1832, labelling this octopus, which he had hunted down in a rock pool on a volcanic island. Then, immersed in seawater, this octopus had been luminous after dark; now it was a furled-up mass of greyish-brown ribbed and ribboned tentacles, swaying slightly in the liquid, distorted by the curve of glass. So there *were* ghosts here in the Spirit Store; and writing history was a little like the work of a taxidermist or zoologist, too, I thought for a moment: working only with preserved relics, the historian was something of an articulator of bones, like Mr Venus in Charles Dickens's *Our Mutual Friend*, putting together the whole creature piece by piece from the jigsawed bones and fragments of dried skin.

I had seen what I needed to see here, but Dr Friday had other

Darwin relics to show me, so I followed him down more corridors and steps that seemed to be getting narrower. Eventually he unlocked another door: the Mollusc Room, he announced. Not his territory – run by the mollusc people – but they had agreed to lend him the key. Here all the specimens were dried and contained in tiny drawers in hundreds of oak cabinets. It was darker in here than it had been in the Spirit Store. We eased our way down into the room between the cabinets, careful not to push at anything, and my guide drew my attention to a smaller oak cabinet for which he had a tiny metal key. Here it was – whatever it was he had to show me. He unlocked the door and opened it.

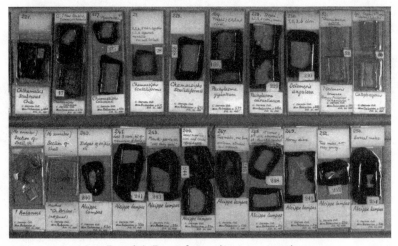

31 Darwin's Barnacles – microscope specimens

Inside, there were twenty or so narrow drawers, each about an inch deep with a tiny handle and each labelled in a small hand in black ink. Dr Friday offered no further explanation, just gestured to me to look more closely. My eye ran down over the labels, squinting slightly in the gloom, but for a few seconds I was unable to take in what the words on the labels signified: *Alcippe*, *Verrucae*, *Scalpellum*, *Ibla*, *Pollicipes*. Then I understood: these were *Darwin's* microscope slides from his barnacle years. Each contained one of his barnacles or a barnacle body part; the Cambridge Zoology Museum had been bequeathed this unique oak-carved curiosity. There must have been nearly three hundred separate slides here – Darwin's slides. Another talisman.

And there was Mr Arthrobalanus himself, granted his Latin name here: *Cryptophialus minutus* – a whole drawerful of him. My guide, delighted with my speechlessness, opened the *Cryptophialus* drawer and placed a slide in my hand. Darwin had himself sealed these two pieces of thin glass together with a ring of dried black asphalt, I remembered; the black circle contained in its centre what the label claimed was a *Cryptophialus* 'mouth part', too small to see with the naked eye.

Confronted by the remains of the microscopic creature that had so confounded Darwin, I was lost for words at the scale of his self-imposed task. Barnacle dissection of the kind he undertook would have been staggeringly difficult – day after day spent with his eye glued to the microscope and his large hands gripping tiny pins, teasing mandibles from other mouth parts in order to prepare a perfect slide; day after day spent following with his eye the thread of an oesophagus or labrum to discover where it led to, note-taking, meticulous systematic comparisons between one thorax and another.[1]

We carried the slides back to the curator's office from the Mollusc Room in a small wooden box in order to examine Mr Arthrobalanus under a microscope. The room was bright, full of books and skulls of mammals in plastic boxes. On the wall, over a row of coffee mugs hanging on pegs, a calendar showed the skeleton of a bat in flight against a black background. Dr Friday lifted out his old Beck microscope from its carrying case and set it up. I was nervous, not sure what to expect. Somehow I wasn't sure I did want to see Darwin's Mr Arthrobalanus after all this time. My guide was not confident that we would see anything at all – after all, it was nearly 150 years since Darwin had prepared them. No microscope specimen could have lasted that long.

There were ten *Cryptophialus* slides in the box, numbered from 255 to 265 (one was listed as missing). According to Darwin's handwritten catalogue and the labels on the slides themselves, there was only one specimen of a *whole* creature; the rest were body parts labelled 'parts of mouth', or 'prehensile antennae', or 'oesophagus showing teeth and mouth'. We laid the ten slides out on the desk under the anglepoise lamp, each with its dark-brown ring of asphalt encircling a mouth or stomach part, too small to see even as a fleck. Some of these rings were now flat circular smudges of brown – the ring had closed inwards. We selected the slide that looked the cleanest of them all and placed it on the microscope stage. We were looking for a labrum, a kind of upper lip. *Cryptophialus* is unique in the barnacle world for the length both of its penis and of its extendible, spoon-shaped upper lip.

Tick, tock – the clock seemed to be louder than it had been. Perhaps it was the suspense of Mr Arthrobalanus's imminent appearance; perhaps it was also the sound of Darwin's impatience pressed here between two slivers of glass – the ghost of a sound, the ghost of perseverance. Under the microscope the labrum shone, a silvery semi-transparent spoon shape against amber – a labrum that Darwin had removed from Mr Arthrobalanus's mouth with patience and with pins. It was exactly the same shape as the drawing George Brettingham Sowerby Jr had prepared for the *Cryptophialus* plate of drawings. Looking at this very same labrum, Darwin had speculated, struggling with words:

We have seen the great lancet-formed appendage of the labrum, literally fringed with fine hairs, can be erected; and I do not doubt that the prey when entangled by the expanded cirri, is borne against this appendage, and is then, by the retraction of the thorax, dragged down its smooth surface to the mouth, where it is seized by the mandible and the maxillae, which lie like a trap at the bottom of an inclined and movable plane.[2]

It was microscope slide number 260, labelled by Darwin in his catalogue as 'two perfect specimens', that I had hopes for; but here it was immediately apparent that the black circle had almost closed. The asphalt had seeped inwards slowly, like imperceptibly moving lava, in the 150 years since Darwin had sealed the slide. Under the microscope, where I had hoped to see two perfect full specimens, there was only a dark-brown mass, at its edges an explosion of brown and gold. But right at the centre, almost entirely engulfed by the amber lava, the tip of a small foot was just visible, a feathery cirrus, like a small black fan, not waving but drowning.

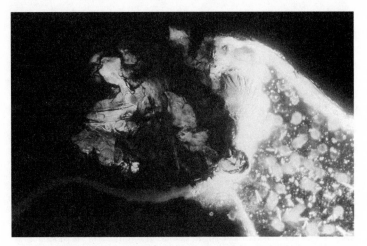

32 Microscopic image of the remains of Mr Arthrobalanus
(*Cryptophialus minutus*)

The asphalt curtain had closed on Mr Arthrobalanus.

Bibliography

*

Ainsworth, W. F., 'Mr Darwin', *Athenaeum*, no. 2846 (1882), 82.

Alborn, Timothy L., 'The Business of Induction: Industry and Genius in the Language of British Scientific Reform, 1820–1840', *History of Science*, vol. 34 (1996), 91–121.

Allen, Mea, *The Hookers of Kew, 1785–1911*, London: Michael Joseph, 1964.

Altick, Richard, *The Shows of London*, Cambridge, Massachusetts, and London: The Belknap Press of Harvard University Press, 1978.

Anderson, D. T., *Barnacles; Structure, Function, Development and Evolution*, London, Glasgow, New York, Tokyo, Melbourne and Madras: Chapman and Hall, 1994.

Appel, Toby A., *The Cuvier-Geoffroy Debate: French Biology in the Decades Before Darwin*. Oxford University Press, 1987.

Armstrong, Patrick, *Darwin's Desolate Islands: A Naturalist in the Falklands 1833 and 1834*, Chippenham: Picton, 1992.

– *Darwin's Other Islands*, University of Durham Research Foundation, 1995.

Ashworth, J. H., 'Charles Darwin as a Student in Edinburgh 1825–1827', *Proceedings of the Royal Society of Edinburgh*, vol. 55 (1935), 97–113.

Atkins, Sir Hedley, *Down: The Home of the Darwins: The Story of A House and the People Who Lived There*, London: Royal College of Surgeons of England, 1974.

Atkinson, H. G. and Martineau, H., *Letters on the Law of Man's Nature and Development*, London: John Chapman, 1851.

Auerbach, Jeffrey A., *The Great Exhibition of 1851: A Nation on Display*, Yale University Press, 1999.

Balfour, John Hutton, *Biography of the Late John Coldstream*, London: James Nisbet and Co., 1865.

Banton, M., ed., *Darwinism and the Study of Society*, London and Chicago: Tavistock Publications and Quadrangle Books, 1961.

Barber, Lynn, *The Heyday of Natural History 1820–1870*, London: Jonathan Cape, 1980.

Barclay, John, *A Series of Engravings Representing the Bones of the Lower Animals*, Edinburgh: E. Mitchell, 1819.

– *An Inquiry into the Opinions, Ancient and Modern, Concerning Life and Organisation*, Edinburgh: Bell and Bradfute, 1822.

{ 262 }

- *Introductory Lectures to a Course of Anatomy with a Memoir by the Author*, Edinburgh: Maclachlan & Stewart, 1827.

Barrett, Andrea, *Voyage of the Narwhal*, London: Flamingo, 1999.

Beauvoir, Simone de, *The Prime of Life*, London: Andre Deutsch and Weidenfield and Nicolson, 1962.

Beck, Richard, *A Treatise on the Construction, Proper Use, and Capabilities of Smith, Beck, and Beck's Achromatic Microscopes*, London: John Van Voorst, 1865.

Beer, Gillian, 'Darwin's Reading and the Fictions of Development', in D. Kohn, ed., *The Darwinian Heritage*, Princeton University Press, 1985.

- 'Four Bodies on the *Beagle*: Touch, Sight and Writing in a Darwin letter', reprinted in Judith Still and Michael Worton, eds., *Textuality and Sexuality. Reading Theories and Practices*, Manchester University Press, 1993.

- *Open Fields: Science in Cultural Encounter*, London: Clarendon Press, 1996.

- *Darwin's Plots: Evolutionary Narrative in Darwin, George Eliot and Nineteenth-Century Fiction*, 2nd edition, Cambridge: CUP, 2000.

Bennett, John Hughes, *Memoir of the Late Professor Edward Forbes*, Edinburgh and London: Sutherland & Knox and Simpkin, Marshall & Co, 1855.

Bennett, William, *Narrative of a Recent Journey of Six Weeks in Ireland*, London: C. Gilpin, 1847, 25-9.

Blackadder, D., 'Mr Blackadder's Account of an Aurora Borealis, observed at Edinburgh, 16th January 1827; with some Particulars of Another, of a Preceding Year', *Edinburgh New Philosophical Journal*, vol. 2, July–September 1827, 342.

Bowlby, John, *Charles Darwin: A Biography*, London: Hutchinson, 1990.

Brightwell, Leonard Robert, *The Zoo Story*, London: Museum Press, London, 1952.

Browne, Janet, *Charles Darwin: Voyaging*, London: Pimlico, 1996.

Bulwer-Lytton, Edward, 'The Confessions of a Water Patient', *New Monthly Magazine*, September 1845.

Burkhardt F. and Smith S., eds., *The Correspondence of Charles Darwin*, 7 vols., Cambridge: CUP, 1985-91.

Cadbury, Deborah, *The Dinosaur Hunters*, London: Fourth Estate, 2000.

Campbell, John, *In Darwin's Wake: Revisiting Beagle's South American Anchorages*, Shrewsbury: Waterline, 1997.

Chambers, Robert, (1844) *Vestiges of the Natural History of Creation and Other Evolutionary Writings*, ed., J. A. Secord, Chicago: Chicago University Press, 1993.

Chesney, Kellow, *The Victorian Underworld*, London: Temple Smith, 1970.

Churchill, F. B., 'Darwin and the Historian', *Biological Journal of the Linnaean Society of London*, vol. 17 (1982,) 45–68.

Clough, A. H., 'Amours de Voyage', in S. Chew, ed., *Arthur Hugh Clough: Selected Poems*, London: Carcanet, 1987.

Coleridge, Samuel Taylor, *Anima Poetae*, London: Heinemann, 1805; 1895.

Colp, Ralph, *To Be an Invalid: The Illness of Charles Darwin*, University of Chicago Press, 1997.

Corbin, Alain, *The Lure of the Sea: The Discovery of the Seaside 1750–1840*, Harmondsworth: Penguin, 1994.

Crisp, D. J., 'Extending Darwin's Investigation on the Barnacle Life-History', *Biological Journal of the Linnean Society*, vol. 20 (1983), 73–83.

Cumming, Roualeyn Gordon, *Five Years of a Hunter's Life in the Far Interior of South Africa*, vols. 1–2, London: John Murray, 1850.

Darwin, Charles, *A Monograph of the Sub-Class Cirripedia, Vol II: The Balanidae*, vol. 13 of *The Works of Charles Darwin*, Paul H. Barrett and R. B. Freeman, eds., London: William Pickering, 1988.

- *Monographs of the Fossil Lepadidae and the Fossil Balanidae*, vol. 14 of *The Works of Charles Darwin*, Paul H. Barrett, and R. B. Freeman, eds., London: William Pickering, 1988.

- *The Voyage of the Beagle*, Harmondsworth: Penguin, 1839; 1989.

- *The Collected Papers of Charles Darwin*, ed. Barrett, University of Chicago Press, 1977.

- *A Monograph on the Sub-Class Cirripedia: The Lepadidae or Pedunculated Cirripedes*, London: Ray Society, 1851.

- *The Autobiography of Charles Darwin 1809–1882*, ed. N. Barlow, London: Collins, 1958.

Darwin, Francis, *The Completed Edited Manuscript of Francis Darwin's Preliminary Draft of the 'Reminiscences of My Father's Everyday Life'*, ed. Robert Brown, manuscript in possession of the Darwin Correspondence Project, Cambridge University Library.

Davies, S., *Emily Brontë: Heretic*, London: The Woman's Press, 1994.

Davis, John, *The Great Exhibition*, Stroud: Sutton Publishing, 1999.

Desmond, Adrian, 'Robert E. Grant: The Social Predicament of a Pre-Darwinian Transmutationist', *The Journal of the History of Biology*, vol. 17, no. 2 (Summer 1984), 189–223.

- *Archetypes and Ancestors. Palaeontology in Victorian London 1850–1875*, London: Blond and Biggs, 1982.

- *The Politics of Evolution: Morphology, Medicine and Reform in Radical London*, University of Chicago Press, 1989.

- *Huxley: From Devil's Disciple to Evolution's High Priest*, Harmondsworth: Penguin, 1998.

Desmond, Adrian and Moore, James, *Darwin*, Harmondsworth: Penguin, 1989.

Desmond, Ray, *Sir Joseph Hooker and India*, London: Linnaean Society, 1993.

- *Sir Joseph Dalton Hooker: Traveller and Plant Collector*, Woodbridge: Antique Collectors Club with the Royal Botanic Gardens, Kew, 1999.

Dickens, Charles, *Bleak House* (1853), Harmondsworth: Penguin, 1986.

Disraeli, B., *Tancred, or the New Crusade* (1847), London: Peter Davies, 1921.

Edwardes, Sir Herbert, *A Year on the Punjab Frontier in 1848–49*, London: Richard Bentley, 1851.

Edwards, John, *London Zoo from Old Photographs 1852–1914*, London: 26 Rhondda Grove, E3 5AP.

Engels, F. and Marx, K., *The Communist Manifesto* (1848), Oxford University Press, 1992.

Ehrenberg, C., 'On the Remarkable Diffusion of Coralline Animalcules from the Use of Chalk in the Arts of Life as Observed by Ehrenberg', *The Annals of National History*, December 1839, 57.

Eiseley, L., *Darwin's Century*, London: Victor Gollancz, 1958.

Endersby, Jim, 'Gentlemanly Generation: Sex, Reproduction and Marriage', in *Cambridge Companion to Darwin*, forthcoming, Cambridge University Press, 2003.

FitzRoy, Capt. Robert, *Narrative of the Surveying Voyages of His Majesty's Ships* Adventure *and* Beagle *between the years 1826 and 1836*, appendix to volume 2, New York: AMS Press, 1966; a reprint of the 1839 edition, by Henry Colburn Publishers, London.

Fleming, John, *The Philosophy of Zoology*, Edinburgh: Archibald Constable, 1822, 40.

Forbes, Edward, *A History of British Starfishes and Other Animals of the Class Echinodermata*, London: John Van Voorst, 1841.

– *A Monograph on the British Naked-Eyed Medusae*, London: Ray Society, 1848.

– 'Shell Fish: Their Ways and Works', *Westminster Review*, vol. 57 (1852), 42–61.

Freeman, R. B., 'Darwin's Negro Bird-Stuffer', *Notes and Records of the Royal Society of London*, vol. 33 (1978–9), 83–6.

Galton, Francis, *The Narrative of an Explorer in Tropical South Africa*, London: John Murray, 1853.

Ghiselin, Michael T., *The Triumph of the Darwinian Method*, University of California Press, 1969.

Ghiselin, M. T. and Jaffe, L., 'Phylogenetic Classification in Darwin's monograph on the subclass Cirripedia', *Systematic Zoology*, vol. 22 (1973), 132–40.

Gosse, P. H., *A Text Book of Zoology for Schools*, London: Society for the Promotion of Christian Knowledge, 1851.

– *A Naturalist's Sojourn in Jamaica*, assisted by Richard Hill, London: Longman, Brown, Green & Longmans, 1851.

– *A Naturalist's Ramble on the Devonshire Coast*, London: John Van Voorst, 1853.

– *The Aquarium: An Unveiling of the Wonders of the Deep Sea*, London: John Van Voorst, 1854.

Gould, S. J., 'Introduction' to *The Mismeasure of Man*, New York: Norton, 1981.

Grant, Robert, 'Observations and Experiments on the Structure and Functions of the Sponge', *Edinburgh Philosophical Journal*, vol. 13, no. 25 (1825), 95–107.

– 'Observations and Experiments on the Structure and Functions of the Sponge', *Edinburgh Philosophical Journal*, vol. 13, no. 26 (1825), 332–46.

– 'Observations and Experiments on the Structure and Functions of the Sponge', *Edinburgh Philosophical Journal*, vol. 14, no. 27 (1826), 113–24.

– 'On the Structure and Nature of the Spongilla friabilis', *Edinburgh Philosophical Journal*, vol. 14, no. 28 (1826), 271–85.

– 'Observations and Experiments on the Structure and Functions of the Sponge', *Edinburgh Philosophical Journal*, vol. 14, no. 29 (1826), 336–41.

– 'Observations on the Structure of Some Silicious Sponges', *Edinburgh New Philosophical Journal*, vol. 1 (1826), 341–51.

– 'Observations on the Structure and Functions of the Sponge', *Edinburgh New Philosophical Journal*, vol. 2 (1826), 121–41.

– 'Observations on the Spontaneous Motions of the Ova of Zoophytes', *Edinburgh New Philosophical Journal*, vol. 2 (1826), 156.

– 'On the Structure and Characters of the Octopus ventricosus, Gr. (Sepia octopodia, Pent.) a Rare Species of Octopus from the Firth of Forth', *Edinburgh New Philosophical Journal*, vol. 3 (1826), 309–17.

Graves, George, *The Naturalist's Companion: Being a Brief Introduction to the Different Branches of Natural Science*, London: Longman, Hurst, Rees, Orme and Brown, 1818.

Green, Toby, *Saddled with Darwin: A Journey through South America*, London: Weidenfeld and Nicolson, 1999.

Grenville, J. A. S., *Europe Reshaped 1848–1878*, London: Fontana, 1976, 30–83.

Grierson, Janet, *Dr Wilson and His Malvern Hydro: Park View in the Water Cure Era*, Malvern: Cora Weaver Press, 1998.

Gully, J., *The Water Cure in Chronic Disease*, London and Malvern: John Churchill and Henry Lamb, 1846.

Haight, Gordon S., *George Eliot & John Chapman with Chapman's Diaries*, New Haven: Yale University Press, 1940.

Hall-Jones, Roger, *A Malvern Bibliography: An Historical and Descriptive Account of the Books Published about Malvern 1725–1987*, Malvern: First Paige, 1988.

Haraway, Donna, *Primate Visions*, New York: Routledge, 1989.

Harrison, J. F. C., *Early Victorian Britain, 1832–51*, London: Fontana, 1988.

Harvey, William Henry, *The Sea Side Book: Being an Introduction to the Natural History of the British Coasts*, London: John Van Voorst, 1849.

Healey, Edna, *Wives of Fame: Jenny Marx, Mary Livingstone, Emma Darwin*, Seven Oaks: New English Library, 1988.

– *Emma Darwin: The Inspirational Wife of a Genius*, London: Headline Book Publishing, 2001.

Herbert, Sandra, ed., *The Red Notebook of Charles Darwin*, London: British Museum, 1980.

Hill, Rowland, *Post Office Reform: Its Importance and Practicability*, London: Charles Knight and Co., 1838.

Hill, Rowland and Hill, George Birbeck, *The Life of Sir Rowland Hill and the History of Penny Postage*, London: De La Rue & Co., 1880.

Holland, Henry, *Chapters on Mental Physiology*, London: Longman, Brown, Green and Longmans, 1852.

Hooker, Joseph Dalton, *The Botany of the Antarctic Voyage of H.M. Discovery Ships Erebus and Terror in the Years 1839–1843 under the Command of Captain Sir James Clark Ross. Part II Flora Novae-Zelandiae*, London: Lovell Reeve, 1853.

Hooker, Joseph Dalton, *Himalayan Journals; or Notes of a Naturalist in Bengal, The Sikkim and Nepal Himalayas, The Khasia Mountains, &c.*, 2 vols., London: John Murray, 1854.

Hooker, Joseph Dalton with Thomas Thomson, *Flora Indica*, Introductory essay, London: W. Pamplin, 1855.

Howitt, William, *The Book of the Seasons; or the Calendar of Nature*, London: Henry Colburn and Richard Bentley, 1831.

Hulme, Peter, *Colonial Encounters. Europe and the Native Caribbean 1492–1797*, London: Routledge, 1986, reprinted 1992.

Huxley, T. H., 'On the Morphology of the Cephalous Mollusca, as illustrated by the anatomy of certain Heteropoda and Pteropoda collected during the voyage of the H.M.S. Rattlesnake in 1846–50', *Philosophical Transactions of the Royal Society of London*, 143: 29–65.

Jacyna, L. S., 'Immanence or Transcendence: Theories of Life and Organisation in Britain, 1790–1835', *Isis*, vol. 74 (1983), 311–29.

Jenyns, Revd Leonard, 'On the Present State of Zoology', *Magazine of Zoology and Botany*, vol. 1 (1837).

Jespersen, Helveg, 'Charles Darwin and Dr Grant', *Lychnos*, 1948–9, 159–67.

Johnston, George, *Terra Lindisfarnensis: Natural History of the Eastern Borders*, London: John Van Voorst, 1853.

Keynes, Randal, *Annie's Box: Charles Darwin, his Daughter and Human Evolution*, London: Fourth Estate, 2001.

Keynes, Richard, *The Beagle Record: Selections from the Original Pictorial Records and Written Accounts of the Voyage of H.M.S Beagle*, Cambridge University Press, 1979.

Keynes, Richard, ed., *Charles Darwin's Beagle Diary*, Cambridge University Press, 1988.

– *Charles Darwin's Zoology Notes and Specimen Lists from HMS Beagle*, Cambridge University Press, 2000.

King-Hele, Desmond, *Erasmus Darwin: A Life of Unequalled Achievement*, London: Giles de la Mare Publishers, 1999.

Kingsley, Charles, *Glaucus; or the Wonders of the Shore*, Cambridge: Macmillan and Co., 1855.

LaBarre, E. J., *Dictionary of Encyclopedia of Paper and Paper-Making*, Amsterdam: Swets and Zeitlinger, 1969.

Lane, R. J., *Life at the Water Cure; or a Month at Malvern: A Diary*, London: John Van Voorst, 1846.

Lee, Charles E., *The Horse Bus As A Vehicle*, London: London Transport, 1974.

Leech, Joseph, *Three Weeks in Wet Sheets, Being the Diary and Doings of a Moist Visitor to Malvern*, London, Bristol and Malvern: Hamilton, Adams and Co., John Ridler, and Lamb and Son, 1856.

Lewes, George Henry, 'Lyell and Owen on Development', *Leader*, 18 October 1851, 996.

Liebig, Justus von, *Familiar Letters On Chemistry, In Its Relations To Physiology, Dietics, Agriculture, Commerce, And Political Economy*, 3rd ed., London: Taylor, Walton & Maberly, 1851.

Litchfield, H. E., ed., *Emma Darwin, Wife of Charles Darwin: A Century of Family Letters*, vols. 1-2. London: John Murray, 1915.

Love, Alan, 'Darwin and Cirripedia Prior to 1846: Exploring the Origins of the Barnacle Research', *Journal of the History of Biology*, vol. 35 (2002), 251-289.

Lurie, E., *Louis Agassiz; A Life in Science*, University of Chicago Press, 1960.

Lynn, C., 'Colours and other Materials of Historic Wallpaper', *Wallpaper Conservation: A Special Issue*, JAIC, vol. 20 (2) (1981), 8-65.

Lyell, Charles, *Principles of Geology* (1830), Harmondsworth: Penguin, 1997.

McNeil, Maureen, *Under the Banner of Science: Erasmus Darwin and His Age*, Manchester: MUP, 1987.

Manier, Edward, *The Young Darwin and his Cultural Circle; A Study of the Influences which Helped Shape the Language and Logic of the First Drafts of the Theory of Natural Selection*, Dortrecht, Holland; Boston, USA: D. Reidel Publishing Company, 1978.

Mantell, Gideon, *The Journal of Gideon Mantell, Surgeon and Geologist, 1818-1852*, London: Oxford University Press, 1940.

Marks, Richard Lee, *Three Men of the Beagle*, New York: Knopf, 1991.

Marshall, James Scott, *The Life and Times of Leith*, Edinburgh: John Donald, 1986.

Martineau, Harriet, *How to Observe: Morals and Manners*, London: John Chapman, 1838.

- *Eastern Life; Present and Past*, 3 vols., London: Edward Moxon, 1848.

Milne-Edwards, H., 'Considérations sur quelques principes relatifs à la classification naturelle des animaux, et plus particulièrement sur la distribution méthodique des mammifères', *Annales des Sciences Naturelles*, 3rd ed., vol. 1 (1844), 65-129.

Moore, James, 'Of Love and Death: Why Darwin "gave up Christianity"', in Moore, J., ed., *History, Humanity and Evolution: Essays for John C. Greene*, Cambridge University Press, 1989.

Moore, James, ed., *History, Humanity and Evolution: Essays for John C. Greene*, Cambridge University Press, 1989.

Moorehead, Alan, *Darwin and the Beagle*, Harmondsworth: Penguin, 1969.

Mowat, Sue, *The Port of Leith: Its History and Its People*, Edinburgh: John Donald in association with the Forth Ports PLC, 1994.

Mundy, Lt. Colonel Godfrey Charles, *Our Antipodes: or, Residence and Ramblers in the Australasian Colonies with a Glimpse of the Gold Fields*, 3 vols., London:

Richard Bentley, 1852.

Newman, Francis, *Phases of Faith; or Passages from the History of My Creed*, London: John Chapman, 1850.

Newman, William A., 'Darwin and Cirripedology', *Crustacean Issues*, vol. 8 (1993), 349–434.

Nicholas, F. W., *Charles Darwin in Australia*, Cambridge University Press, 1989.

Nott, James, *The Story of the Water Cure as Originated at Graefenberg and Perfected at Malvern*, Malvern: Stevens & Co., 1990.

Nyart, Lynn, *Biology Takes Form: Animal Morphology and the German Universities 1800–1900*, Chicago and London: University of Chicago Press, 1995.

Odescalchi, Elani, *Charles Dickens at Malvern*, Malvern: First Paige, 1992.

Ospovat, Dov, *The Development of Darwin's Theory: Natural History, Natural Theology and Natural Selection, 1838–1859*, Cambridge University Press, 1981.

Owen, R., *Lectures on the Comparative Anatomy and Physiology of the Invertebrate Animals*, London: Longman, Brown, Green and Longmans, 1843.

Phillips, Adam, *Darwin's Worms*, London: Faber and Faber, 1999.

Porter, Duncan, 'The Beagle Collector and His Collections', in D. Kohn, ed., *The Darwinian Heritage*, Princeton University Press, 1985.

Porter, Duncan, 'On the road to the *Origin* with Darwin, Hooker, and Gray', *Journal of the History of Biology*, vol. 26 (1993), 1–38.

Porter, Roy, *The Medical History of Waters and Spas*, London: Wellcome Institute for the History of Medicine, 1990.

– *The Greatest Benefit to Mankind: A Medical History of Humanity from Antiquity to the Present*, London: Fontana, 1997.

Price, R., *The Revolutions of 1848*, Basingstoke: Macmillan, 1988.

Raff, R. A. & Kaufman T. C., *Embryos, Genes and Evolution: The Developmental-Genetic Basis of Evolutionary Change,* New York: Macmillan Publishing Co., 1983.

Ray, Desmond, *Sir Joseph Hooker and India*, London: Linnaean Society, 1993.

– *Sir Joseph Dalton Hooker: Traveller and Plant Collector*, Antique Collectors Club, 1999.

Reed, John, *London Buses Past and Present*, London: Capital Transport, 1988.

Rehbock, F., *The Philosophical Naturalists: Themes in Early Nineteenth-Century British Biology*, Madison: University of Wisconsin Press, 1983.

– 'The Early Dredgers: Naturalising in British Seas, 1830–1850', *Journal of History of Biology*, vol. 12 (1979), 293–368.

– 'The Victorian Aquarium in Ecological and Social Perspective', in M. Sears and D Merriman, eds., *Oceanography: The Past*, New York: Springer-Verlag, 1980.

Richmond, Marsha, 'Darwin's Study of the Cirripedia', in Burkhardt F. and Smith S., eds., *The Correspondence of Charles Darwin*, 7 vols., Cambridge: CUP, 1985–91; vol. 4, 388–409.

Rylance, R., *Victorian Psychology and British Culture 1850–1880*, Oxford: Oxford University Press, 2000.

Secord, James A., 'Edinburgh Lamarckians: Robert Jameson and Robert E. Grant', *Journal of History of Biology*, vol. 24 (1991), 1–18.

– *Victorian Sensation: The Extraordinary Publication, Reception, and Secret Authorship of Vestiges of the Natural History of Creation*, Chicago: University of Chicago Press, 2000.

[Sedgwick, Adam], 'Natural History of Creation', *Edinburgh Review*, vol. 82 (July 1845), 1–85.

Sheppersen, George, 'The Intellectual Background of Charles Darwin's Student Years at Edinburgh', in M. Banton, ed., *Darwinism and the Study of Society*, London and Chicago: Tavistock Publications and Quadrangle Books, 1961, 17–35.

Simmons, J. and Biddle, G., eds., *The Oxford Companion to Railway History from 1603 to the 1990s*, Oxford University Press, 1997.

Sloan, Phillip R., 'Darwin's Invertebrate Program, 1826–1836: Preconditions for Transformism', in D. Kohn, ed., *The Darwinian Heritage*, Princeton University Press, 1985.

– 'Darwin, Vital Matter and the Transformism of Species', *Journal of the History of Biology*, vol. 19 (1986), 369–445.

Smith, Jonathan, *Fact and Feeling: Baconian Science and the Nineteenth-Century Literary Imagination*, Madison: University of Wisconsin Press, 1994.

Southward, A. J., 'A New Look at Variation in Darwin's Species of Acorn Barnacles', *Biological Journal of the Linnaean Society*, vol. 20 (1983), 59–72.

Sperber, J., *The European Revolutions, 1848–1851*, Cambridge University Press, 1994.

Stanbury, D., ed., *A Narrative of the Voyage of the HMS Beagle*, London: Folio Society, 1977.

Stott, Rebecca, 'Thomas Carlyle and the Crowd Revolution, Geology and the Convulsive "Nature" of Time', *Journal of Victorian Culture*, 4.1 (Spring 1999), 1–24.

– 'Darwin's Barnacles: Mid-Century Victorian Natural History and the Marine Grotesque', in R. Luckhurst and J. McDonagh, *Transactions and Encounters: Science and Culture in the Nineteenth Century*, Manchester: Manchester University Press, 2002, 151–181.

Street, Philip Arthur Richard, *The London Zoo*, London: Odhams Press, 1956.

Sulivan, Henry Norton., ed., *Life and Letters of the Late Admiral Sir Batholomew James Sulivan*, London: John Murray, 1896.

Sulloway, Frank, 'Darwin's Conversion: The *Beagle* Voyage and Its Aftermath', *Journal of the History of Biology*, vol. 15 (1982), 1–53.

Syme, Patrick, ed., *Werner's Nomenclature of Colours with Additions*, 2nd ed., Edinburgh and London: William Blackwood and T. Cadell, 1821.

Thompson, John Vaughan, *Zoological Researches and Illustrations; or, Natural History of Nondescript or Imperfectly Known Animals in a Series of Memoirs*, Cork: King and Ridings, 1830.

– *The Pestilential Cholera Unmasked*, Cork: King and Ridings, 1832.

Thwaite, Ann, *Glimpses of the Wonderful: The Life of Philip Henry Gosse*, London: Faber and Faber, 2002.

Thynne, Anna, 'On the Increase of Madrepores', *The Annals and Magazine of Natural History*, no. 18 (June 1859), 450.

Timbs, John, *Curiosities of London: Exhibiting the Most Rare and Remarkable Objects of Interest in the Metropolis with Nearly Fifty Years' Personal Recollections*, London: David Bogue, 1855.

Trenn, T., 'Charles Darwin, Fossil Cirripeds and Robert Fitch: Presenting Sixteen Hitherto Unpublished Darwin Letters of 1849 to 1851', *Proceedings of the American Philosophical Society*, vol. 118 (1974), 471–91.

Trotter, David, *Circulation: Defoe, Dickens and the Economies of the Novel*, Basingstoke: Macmillan, 1988.

Turner, E. S., *Taking the Cure*, London: Michael Joseph, 1967.

Wakley, Thomas, 'Biographical Sketch of Robert Edmund Grant, M.D.', *The Lancet*, vol. 2 (1850), 686–95.

Wallace, Joyce M., *Traditions of Trinity and Leith*, Edinburgh: John Donald, 1997.

Walvin, James, *Beside the Seaside: A Social History of the Popular Seaside Holiday*, London: Allen Lane, 1978.

Weaver, Cora, *A Short Guide to Charles Darwin and Evelyn Waugh in Malvern*, Malvern: Cora Weaver Press, 1991.

– *A Short Guide to Malvern as a Spa Town*, Malvern: Cora Weaver Press, 1991.

– *A Short Guide to Elizabeth Barrett and George Bernard Shaw in Malvern*, Malvern: Cora Weaver Press, 1992.

Weinshank, Donald J., *A Concordance to Charles Darwin's Notebook 1836–1844*, Ithaca: Cornell University Press, 1990.

Whewell, William, 'Review of *The Works of Francis Bacon*, ed. James Spedding, Robert Leslie Ellis and Douglas Demon Heath', *Edinburgh Review*, 106 (1857), 314–15.

Winsor, Mary, *Starfish, Jellyfish and the Order of Life: Issues in Nineteenth-Century Science*, New Haven: Yale University Press, 1969.

– 'Barnacle Larvae in the Nineteenth Century: A Case Study in Taxonomic Theory', *Journal of the History of Medicine and Allied Sciences*, vol. 24 (1969), 294–309.

Winton, Keith, *Murders in Edinburgh*, Edinburgh: Edinburgh Impressions, 1985.

Wood, R. D., 'The Diorama in Great Britain in the 1820s', *The History of Photography*, vol. 17, no. 13 (Autumn 1993), 284–95.

Yeo, Richard, 'An Idol in the Market-Place: Baconianism in Nineteenth-Century Britain', *History of Science*, vol. 23 (1985), 251–98.

– *Defining Science: William Whewell, Natural Knowledge and Public Debate in Early Victorian Britain*, Cambridge University Press, 1993.

Young, Alex, *The Encyclopaedia of Scottish Executions 1750 to 1963*, Kent: Eric Dobby Publishing, 1998.

Journals

Annals and Magazine of Natural History (London, 1841–1966)
Athenaeum (London, 1830–70)
Calcutta Journal of Natural History (Calcutta, 1840–47)
Illustrated London News (London, 1842–present)
London's Magazine of Natural History (London, 1829–40)
Magazine of Zoology and Botany (Edinburgh, 1837–38)
Titan (London and Edinburgh, 1856–59)
Water-Cure Journal and Hygienic Magazine (London, 1848–50)
Zoologist (London, 1843–1916)

Endnotes

*

1 The Sponge Doctor

1 The *Annual Register*, vol. 64 (1822), p. 10, records that the weather was
exceptionally mild in January, particularly in Scotland, where summer flowers
were recorded as blooming in Edinburgh gardens. For information about the
execution of the pirates, see Alex Young (1998), *The Encyclopaedia of Scottish
Executions 1750 to 1963*, Eric Dobby Publishing, Kent.

2 See Janet Browne, *Charles Darwin: Voyaging* (London: Pimlico, 1995), p. 58–9.

3 See painting of Leith Races by William Reed, City of Edinburgh Art Centre.

4 He joined the Medico-Chirurgical Society at the age of 18 and was elected
President within months. At 19 he joined the Royal Medical Society and was
elected their President three years later.

5 For further information on Professor Jameson see James A. Secord,
'Edinburgh Lamarckians: Robert Jameson and Robert E. Grant', *Journal of
History of Biology*, vol. 24 (1991), pp. 1–18.

6 Erasmus Darwin, *Zoonomia, or the Laws of Organic Life*, Dublin (printed for
P. Byrne, and W. Jones, 1794–6), 2 vols, vol. 1, p. 506.

7 Ibid., p. 509.

8 For work on Erasmus Darwin see Desmond King-Hele, *Erasmus Darwin: A
Life of Unequalled Achievement* (London: Giles de la Mare Publishers, 1999)
and Maureen McNeil, *Under the Banner of Science: Erasmus Darwin and His
Age* (Manchester: MUP, 1987).

9 See Roy Porter, *The Greatest Benefit to Mankind: A Medical History of
Humanity From Antiquity to the Present* (London: Fontana, 1997), pp. 306–14.

10 See Phillip R. Sloan, 'Darwin's Invertebrate Program, 1826–1836:
Preconditions for Transformism' in D. Kohn, ed. (1985), *The Darwinian
Heritage* (Princeton: Princeton University Press, 1985), p. 75.

11 The most detailed research into Grant's life and work has been undertaken by
Adrian Desmond in *The Politics of Evolution: Morphology, Medicine and Reform
in Radical London* (Chicago: University of Chicago Press, 1989), 'Robert E.
Grant's Later Views on Organic Development', *ANH*, vol. 11 (1984), pp. 395–413
and 'Robert E. Grant: The Social Predicament of a Pre-
Darwinian Transmutationist', *Journal of the History of Biology*, 17: 2 (Summer
1984), pp. 189–223. Grant, trained in Parisian medical research techniques,
took notes about everything he dissected, all his ideas, particular trains of

thought and critical conversations. But all of this has disappeared. He died unmarried with no close relatives and, although his library survived, these valuable journals and bundles of notes and letters did not. So tracing his intellectual and physical journeys is a matter of detective work. There is a short biographical essay written by his friend Thomas Wakley in 1850 for *The Lancet* and probably based on interviews with Grant, but almost nothing else, apart from the dozens of essays he published in the 1820s and 1830s. Thomas Wakley, 'Biographical Sketch of Robert Edmund Grant, M.D.', *The Lancet*, vol. 2 (1850), pp. 686–95.

12 John Barclay, *An Inquiry into the Opinions, Ancient and Modern, Concerning Life and Organisation* (Edinburgh: Bell and Bradfute, 1822), p. 525.

13 Robert Grant, 'Observations and Experiments on the Structure and Functions of the Sponge', *Edinburgh Philosophical Journal*, vol. 14 (1826), no. 27, p. 122.

14 Robert Grant, 'Observations and Experiments on the Structure and Functions of the Sponge', *Edinburgh Philosophical Journal*, vol. 13 (1825), no. 25, p. 97.

15 Ibid., p. 99.

16 Ibid., p. 102.

17 Robert Grant, 'Observations and Experiments on the Structure and Functions of the Sponge', *Edinburgh New Philosophical Journal*, vol. 2 (1826), p. 126.

18 Robert Grant, 'Observations and Experiments on the Structure and Functions of the Sponge', *Edinburgh Philosophical Journal*, vol. 14 (1826), no. 27, p. 123.

19 Robert Grant, 'Observations and Experiments on the Structure and Functions of the Sponge', *Edinburgh New Philosophical Journal*, vol. 2 (1826), p. 136.

20 I am grateful to Adrian Desmond for generous assistance and advice on this matter.

21 For the social history of Leith see James Scott Marshall, *The Life and Times of Leith* (Edinburgh: John Donald, 1986); Sue Mowat, *The Port of Leith: Its History and Its People* (Edinburgh: John Donald in association with the Forth Ports PLC, 1994); Joyce M. Wallace, *Traditions of Trinity and Leith* (Edinburgh: John Donald, 1997).

22 Darwin's 1826 Notebook, DAR 129.

2 Riddles of the Rock Pools

1 CD to Robert W. Darwin, 23 October 1825, *Correspondence* 1, pp. 18– 19. For an excellent account of Edinburgh medical training in this decade see Janet Browne, *Charles Darwin: Voyaging* (London: Pimlico, 1996), chapter 2.

2 CD to Caroline Darwin, January 1825, *Correspondence* 1: p. 25.

3 Susan Darwin to CD, 27 March 1826, *Correspondence* 1: p. 41.

4 *Autobiography*, p. 25.

5 R. B. Freeman, 'Darwin's Negro Bird-Stuffer', *Notes and Records of the Royal*

Society of London, vol. 33 (1978–9), pp. 83–6.

6 Fleming (1822), p. 40.

7 This section is based upon Darwin's Edinburgh Zoology journal: DAR 129.

8 John Hutton Balfour, *Biography of the Late John Coldstream* (London: James Nisbet and Co, 1865), pp. 2–3.

9 Balfour (1865), p. 6.

10 The first historian to suggest that Robert Grant might have been homosexual was Adrian Desmond in *Archetypes and Ancestors: Palaeontology in Victorian London 1850–1875* (London: Blond and Biggs, 1982). It is still conjectural but based upon accounts of Grant's reputation from the Zoology Department at University College London. Grant never married and continued to have intense friendships with men throughout his life; many of them he travelled with for many months abroad. If this were indeed true and if sexual feelings occurred between Coldstream and Grant, this may account for the level of self-loathing that Coldstream expressed in his diaries during the time he worked alongside Grant, particularly given the intensity of Coldstream's religious beliefs. It may also, as Desmond points out, be one possible factor among many in the decline of Grant's reputation in London and Charles Darwin's eventual distancing from him. Again I am grateful to Adrian Desmond for discussions on this matter.

11 See J. H. Ashworth, 'Charles Darwin as a Student in Edinburgh 1825–1827' in *Proceedings of the Royal Society of Edinburgh*, vol. 55 (1935), pp. 97–113. For this period of Darwin's life see the fascinating account by Adrian Desmond and James Moore, *Darwin* (Harmondsworth: Penguin, 1992); Helveg Jespersen, 'Charles Darwin and Dr Grant', *Lychnos (1948–9)*, pp. 159–67; George Sheppersen, 'The Intellectual Background of Charles Darwin's Student Years at Edinburgh', in M. Banton, ed., *Darwinism and the Study of Society* (London: Tavistock Publications; Chicago: Quadrangle Books, 1961), pp. 17–35.

12 *Plinian Minutes MSS*, 1:ff. 34–6, Edinburgh University Library, Dc.2.53.

13 *Autobiography*, p. 48.

14 See Sara Stevenson, *Hill and Adamson's The Fishermen and Women of the Firth of Forth* (Scottish National Portrait Gallery, 1991) and Alain Corbin, *The Lure of the Sea: The Discovery of the Seaside, 1750–1840* (Harmondsworth: Penguin, 1995).

15 D. Blackadder (1827), 'Mr Blackadder's Account of an *Aurora Borealis*, observed at Edinburgh 16th January 1827; with some particulars of another, of a preceding year', *Edinburgh New Philosophical Journal*, vol. 2 (July–September), p. 342.

16 W. F. Ainsworth (1882), 'Mr Darwin', *Athenaeum*, no. 2846, p. 82.

17 DAR 5:28–39.

18 Robert Grant, (1826), 'Observations on the Spontaneous Motions of the Ova of Zoophytes', *Edinburgh New Philosophical Journal*, vol. 2, p. 156.

19 DAR 118: pp. 5–6.

20 Ibid.

21 Cited in J. H. Ashworth (1934–5), 'Charles Darwin as a Student in Edinburgh, 1825–27', *Proceedings of the Royal Society of Edinburgh*, vol. 55, part 2, p. 105.

22 P. Helveg Jespersen (1948–9), 'Charles Darwin and Dr Grant', in *Lychnos*, p. 164–5. The scrap that Jespersen refers too has now been lost, so one should read this reminiscence warily. However, later when Grant worked in London he was notorious for his priority disputes and guarded about his research findings; see Desmond, Adrian, *The Politics of Evolution: Morphology, Medicine and Reform in Radical London* (Chicago: University of Chicago Press, 1989).

23 Balfour (1865), p. 38.

24 Keith Winton (1985), *Murder in Edinburgh* (Edinburgh Impressions), pp. 95–105 and Alex Young (1998), *The Encyclopaedia of Scottish Executions 1750–1963* (Kent: Eric Dobby Publishing, 1985), pp. 102-3.

25 Balfour (1865), p. 69.

26 Letter from John Coldstream to Charles Darwin, 28 February 1829, *Correspondence*, vol. 10, p. 77.

27 Ibid.

28 Cited in Adrian Desmond and James Moore, *Darwin* (Harmondsworth: Penguin, 1991), p. 94.

29 *Autobiography*, pp. 71–2.

3 A Baron Münchausen Amongst Naturalists

1 CD to Caroline Darwin, 28 April 1831, *Correspondence* 1: pp. 125–6.

2 *Diary*, p. 19.

3 See Adrian Desmond and James Moore, *Darwin* (Harmondsworth: Penguin, 1991), p. 116.

4 *Diary*, p. 20.

5 Capt. Robert FitzRoy, *Narrative of the Surveying Voyages of His Majesty's Ships* Adventure *and* Beagle *between the years 1826 and 1836*, appendix to volume 2 (New York: AMS Press, 1966; a reprint of the 1839 edition, by Henry Colburn Publishers, London), p. 92.

6 David Stanbury, ed., *A Narrative of the Voyage of HMS* Beagle (London, Folio, 1977), p. 69.

7 'My father used to say that it was the absolute necessity of tidiness in the cramped space on the Beagle that helped to give him his methodical habits of working.' On the *Beagle*, too, he would say, that he learned what he considered the golden rule for saving time; i.e., 'taking care of the minutes'. Francis Darwin, ed., *The Autobiography of Charles Darwin and Selected Letters* (New York: Dover Publications, 1958; a reprint of the 1892 edition by D. Appleton Publishers, New York), p. 132.

8 CD to Robert W. Darwin, 8 February 1832, *Correspondence* 1, p. 202.

9 DAR 30:1, p. 1.

10 Christain Gottfried Ehrenberg (1835), 'Über die Akalephen des rothen Meeres und den Organismus der Medusen der Ostee,' *Abh. Königl. Akad. Wiss. Berlin*, (Jerg. 1835), 181–260; 233.

11 *Diary*, p. 31.

12 Ibid., p. 35.

13 Ibid., p. 21.

14 J. S. Henslow to CD, 15 January 1833, *Correspondence* 1: p. 293.

15 CD to Catherine Darwin, 14 July 1833, *Correspondence* 1: p. 314.

16 *Diary*, p. 22.

17 Ibid., pp. 30–31.

18 Patrick Syme, ed., *Werner's Nomenclature of Colours with Additions, arranged as to render it highly useful to the arts and sciences* . . . (2 nd ed., Edinburgh, 1821).

19 *Diary*, p. 24.

20 CD to J. S. Henslow, 23 July 1832, *Correspondence* 1: p. 251.

21 *Diary*, p. 36. See also Terry Gilliam's film *The Adventures of Baron Münchausen*.

22 J. H. Balfour, *Biography of the Late John Coldstream* (London: John Nisbet & Co, 1964), p. 94.

23 Ibid., p. 97.

24 Susan Darwin to CD, 18 August 1832, *Correspondence* 1: p. 257.

25 John Thompson, *Zoological Researches and Illustrations; Or Natural History of Nondescript or Imperfectly Known Animals* (Cork: King and Ridings, 1830), p. 40.

26 *Diary*, p. 111.

27 Ibid., p. 22.

28 John Thompson (1830), *Zoological Researches*, pp. 49–50.

29 Charles Darwin, *Voyage of the Beagle* (Harmondsworth: Penguin, 1989), p. 55.

30 See Philip Sloan's interesting breakdown of the notebooks in Philip R. Sloan (1985), 'Darwin's Invertebrate Program, 1826–1836: Preconditions for Transformism' in D. Kohn, ed. (1985) *The Darwinian Heritage* (Princeton: Princeton University Press), p. 89–91. For further information on Charles Lyell see introductory essay by James A. Secord in Charles Lyell, *Principles of Geology* (Harmondsworth: Penguin, 1997).

31 Darwin to W. D. Fox, 23 May 1833, *Correspondence* 1: p. 316.

32 Richard Keynes, ed. (2000), *Charles Darwin's Zoology Notes & Specimen Lists from H.M.S.* Beagle (Cambridge: CUP), p. xi.

33 CD to J. S. Henslow, 23 July 1832, *Correspondence* 1: p. 251.

34 Philip R. Sloan, 'Darwin's Invertebrate Program, 1826–1836: Preconditions for Transformism', in D. Kohn, ed., *The Darwinian Heritage* (Princeton: Princeton University Press, 1985), p. 110.

35 Ibid., p. 111.

36 Charles Darwin, *Voyage of the Beagle* (Harmondsworth: Penguin, 1989),
    p. 202. It is tempting to see in this extract an early version of Darwin's famous
    entangled bank analogy in the *Origin*.

37 See *Diary*, p. 146.

38 Charles Darwin, *Voyage of the Beagle* (Harmondsworth: Penguin, 1989),
    p. 203. The supposed cannibalism of 'barbarian' tribes was one of the
    recurring preoccupations of travel writers in the eighteenth and nineteenth
    centuries. Peter Hulme has shown how often it recurs in colonial narratives
    and has also shown how little actual evidence there was to support the
    supposed extent of cannibalism in the 'uncivilized' world; see Peter Hulme,
    Colonial Encounters. Europe and the Native Caribbean 1492–1797 (London.
    Routledge. 1986, reprinted 1992).

39 *Diary*, p. 222.

40 Charles Darwin, *Voyage of the Beagle* (Harmondsworth: Penguin, 1989), p. 178.

41 Ibid., p. 179.

42 For more detailed accounts of the Fuegians on the *Beagle* and the outcome of
    FitzRoy's experiment see Janet Browne, *Charles Darwin: Voyaging* (London:
    Pimlico, 1996); Adrian Desmond and James Moore, *Darwin*
    (Harmondsworth: Penguin, 1991); Gillian Beer, 'Four Bodies on the *Beagle*:
    touch, sight and writing in a Darwin letter', reprinted in Judith Still and
    Michael Worton, eds., *Textuality and Sexuality. Reading Theories and
    Practices* (Manchester University Press, 1993); and Richard Lee Marks, *Three
    Men of the Beagle* (New York: Knopf, 1991).

43 CD to Susan Darwin, 14 July 1832, *Correspondence* 1: p. 248.

44 CD to Frederick Watkins, 18 August 1832, *Correspondence* 1: p. 261.

45 *Diary*, p. 367.

46 CD to Susan Darwin, 14 July–7 August, 1832, *Correspondence* 1: p. 248.

47 David Stanbury, ed., *A Narrative of the Voyage of the* Beagle (London: Folio,
    1977), p. 222.

48 *Diary*, p. 271–2.

49 David Stanbury, ed., *A Narrative of the Voyage of the Beagle* (London: Folio,
    1977), p. 223.

50 Charles Darwin, *Voyage of the Beagle* (Harmondsworth: Penguin, 1989), p. 338.

51 CD to Catherine Darwin, 20 July 1834, *Correspondence* 1: p. 391.

52 Charles Dickens, *Bleak House* (1853), (Harmondsworth: Penguin, 1986), p. 49.

53 Adrian Desmond and James Moore, *Darwin* (Harmondsworth: Penguin,
    1991), p. 189.

54 Philip R. Sloan, 'Darwin's Invertebrate Program, 1826–1836: Preconditions
    for Transformism', in D. Kohn, ed., *The Darwinian Heritage* (Princeton:
    Princeton University Press, 1985), p. 111.

55 Revd Leonard Jenyns, 'On the Present State of Zoology', *Magazine of Zoology
    and Botany*, vol. 1 (1837), p. 28.

4 Settling Down

1 In the 1840s the village of Down in Kent changed its name to 'Downe' so that it would not be confused with County Down in Ireland. Darwin and his family, however, continued to call their house Down House. To avoid confusion for the reader I will call both the village and the house Down, rather than Downe.

2 Darwin's son Francis wrote in later life that he remembered the study shelves being full of 'odds & ends, glasses, saucers, biscuit boxes, zinc labels, bits of wood, flower pot saucers of sand, and spirits of wine glycerine.' Francis Darwin, *The Completed Edited Manuscript of Francis Darwin's Preliminary Draft of the 'Reminiscences of My Father's Everyday Life'*, ed. Robert Brown, manuscript in possession of the Darwin Correspondence Project, Cambridge University Library, p. 58.

3 Francis Darwin manuscript of *Reminiscences*, p. 56.

4 CD to Robert FitzRoy, 1 October 1846, *Correspondence* 3: p. 345.

5 See Adrian Desmond and James Moore, *Darwin* (Harmondsworth: Penguin, 1992), p. 336.

6 CD to W. D. Fox, 25 March 1843, *Correspondence* 2: p. 352.

7 CD to Emma Darwin, 25 June 1846, *Correspondence* 3: p. 326.

8 CD to Emma Darwin, 24 June, 1846, Correspondence 3: p. 325.

9 Adrian Desmond and James Moore, *Darwin* (Harmondsworth: Penguin, 1992), p. 271.

10 'Darwin's Notes on Marriage: Second Note, July 1838', *Correspondence* 2: pp. 444-5.

11 Ibid.

12 Charles Darwin described his son William as an 'animalcule' in a letter to T. C. Eyton, 6 January 1840, and to Robert FitzRoy, 20 February 1840, *Correspondence* 2: p. 255.

13 *Correspondence* 7: Supplement.

14 Cited in Edna Healey, *Wives of Fame: Jenny Marx, Mary Livingstone, Emma Darwin* (London: Hodder and Stoughton, 1988), p. 241.

15 'On the Remarkable Diffusion of Coralline Animalcules from the Use of Chalk in the Arts of Life as Observed by Ehrenberg', *The Annals of Natural History*, December 1839, p. 57; see also C. Lynn, 'Colors and other Materials of Historic Wallpaper', in *Wallpaper Conservation: A Special Issue. JAIC*, vol. 20, no. 2, (1981) pp. 58-65 and E. J. LaBarre, *Dictionary and Encyclopaedia of Paper and Paper-making* (Amsterdam: Swets & Zeitlinger, 1969).

16 For further information on Joseph Hooker see Ray Desmond, *Sir Joseph Dalton Hooker* (Antique Collectors Club, 1999) and a forthcoming essay by Jim Endersby, 'Gentlemanly Generation: Darwin on Heredity, Reproduction and Marriage', *Cambridge Companion to Darwin* (Cambridge: CUP, 2002).

17 CD to J. D. Hooker, 11 January, 1844, *Correspondence* 3: p. 2.

18 CD to Leonard Jenyns, 12 October 1844, Darwin, *Correspondence* 3: p. 67–8.

19 For a splendid and comprehensive account of the history and reception of *Vestiges* read James Secord, *Victorian Sensation: The Extraordinary Publication, Reception and Secret Authorship of 'Vestiges of the Natural History of Creation'* (Chicago and London: The University of Chicago Press, 2000).

20 See Benjamin Disraeli, *Tancred; or, the New Crusade* (1847; London: Peter Davies, 1927), pp. 112–13.

21 CD to Hooker, 7 January 1845, *Correspondence* 3: p. 108.

22 [Sedgwick] 'Natural History of Creation' *Edinburgh Review* (1845); see also discussion of Segwick's reaction in James Secord, *Victorian Sensation: The Extraordinary Publication, Reception and Secret Authorship of 'Vestiges of the Natural History of Creation'* (Chicago and London: The University of Chicago Press, 2000), pp. 231–47.

23 CD to J. D. Hooker, 10 September 1845, *Correspondence* 3: p. 253.

24 Darwin, Notebook C: 76–7.

25 Letter to J. D. Hooker, 2 October 1846, *Correspondence* 3: p. 346.

26 See Charles E. Lee (1974), *The Horse Bus As A Vehicle* (London: London Transport, 1974); John Reed (1988), *London Buses Past And Present* (London: Capital Transport, 1988); also the website of the London Transport Museum at http://www.ltmuseum.co.uk

27 CD to J. D. Hooker, 12 November 1846, *Correspondence* 3: p. 358, note 7.

28 CD to J. D. Hooker, 26 October 1846, *Correspondence* 3: p. 357.

29 Ibid., p. 357.

30 CD to J. D. Hooker, 12 November 1846, *Correspondence* 3: p. 365.

31 Ibid.

32 Bill Newman has argued in 'Darwin and Cirripedology', in *Crustacean Issues*, vol. 8 (1993), p. 379 that this double penis was in fact 'two median-dorsal filaments' on the female of Cryptophialus; but as Darwin assumed all barnacles to be hermaphrodite, it did not occur to him as yet that separate sexes might be present in some sub-groups.

33 CD to Sir James Ross, 31 December 1847, *Correspondence* 4: p. 101.

34 See Rowland Hill and George Birkbeck Hill, *The Life of Sir Rowland Hill and the History of Penny Postage*, 2 vols (London: De La Rue & Co, 1880) and Rowland Hill, *Post Office Reform: Its Importance and Practicability* (London: C. Knight, 1837).

35 J. F. C. Harrison, *Early Victorian Britain, 1832–51* (London: Fontana, 1988), p. 25.

36 Kellow Chesney, *The Victorian Underworld* (London: Temple Smith, 1970), p. 34.
    See also J. F. C. Harrison, *Early Victorian Britain, 1832–51* (London: Fontana, 1988), p. 49–50.

37 See James Walvin, *Beside the Seaside: A Social History of the Popular Seaside Holiday* (London: Allen Lane, 1979), pp. 36–71.

38 Charles Darwin, 'Autobiographical Fragment', *Correspondence* 2: p. 440.

39 Edward Forbes, *A Monograph of the British Naked-Eyed Medusae* (London: Ray Society, 1848), p. 60.

40 Adrian Desmond, *Huxley: From Devil's Disciple to Evolution's High Priest* (Harmondsworth: Penguin, 1998), p. 70.

5 Better Than Castle-Building

1 In a letter to J. E. Gray, Keeper of the Zoological Department, British Museum, 18 December 1847, Darwin says: 'it is my intention to publish a monograph on this difficult order', *Correspondence* 4: p. 99.

2 Revd Leonard Jenyns, 'On the Present State of Zoology', *Magazine of Zoology and Botany*, vol. 1 (1837), p. 28.

3 Quoted by Darwin in letter to J. S. Henslow, 1 April 1848, *Correspondence* 4: p. 128.

4 *The Times*, Leading Article, 25 February 1848, p. 5, col. a.

5 See J. A. S. Grenville, *Europe Reshaped 1848–1878* (London: Fontana, 1976), p. 30.

6 *The Times*, 4 March 1848, p. 4., col. e.

7 Grenville, op. cit., p. 35.

8 Cited in Stevie Davies, *Emily Brontë: Heretic* (London: The Women's Press, 1994), p. 245. For a study of the impact of geological ideas on the understanding of the nature of revolution in the nineteenth century see Rebecca Stott, 'Thomas Carlyle and the Crowd: Revolution, Geology and the Convulsive "Nature" of Time', in *Journal of Victorian Culture* vol. 4, no. 1 (Spring 1999), pp. 1–24.

9 From Arthur Hugh Clough, 'Amours de Voyage', in Shirley Chew, ed., *Arthur Hugh Clough: Selected Poems* (London: Carcanet, 1987), p. 142. For further information on the revolutions of 1848 see Jonathan Sperber, *The European Revolutions, 1848–1851* (Cambridge: Cambridge University Press, 1994); R. Price, *The Revolutions of 1848* (Basingstoke: Macmillan, 1988). For an analysis of how the Victorians described and imagined revolution in geological terms see my essay 'Thomas Carlyle and the Crowd: Revolution, Geology and the Convulsive "Nature" of Time', in *The Journal of Victorian Culture*, vol. 4, no. 1 (Spring 1999), pp. 1–24.

10 I am grateful to Dr Adrian Friday, University Lecturer and Curator of Vertebrates at Cambridge Zoology Museum, for this information about the dissection of barnacles. He tells me that barnacle dissection is known to make or break first-year zoology undergraduates, particularly the dissection of coned barnacles.

11 CD to J. S. Henslow, 1 April 1848, *Correspondence* 4: p. 128.

12 Charles Darwin, *A Monograph on the Sub-Class Cirripedia: The Lepadidae or Pedunculated Cirripedes* (London: Ray Society, 1851), p. 293.

13 Ibid., p. 214.

14 See 'Darwin and the Branching Conception' in Dov Ospovat, *The Development of Darwin's Theory: Natural History, Natural Theology and Natural Selection, 1838–1859* (Cambridge and New York: CUP, 1981), pp. 146–69.

15 Philip R. Sloan, 'Darwin's Invertebrate Program, 1826–1836: Preconditions for Transformism', in D. Kohn, ed., *The Darwinian Heritage* (Princeton: Princeton University Press, 1985), p. 111.

16 Darwin, letter to J. S. Henslow, 1 April 1848, *Correspondence* 4: p. 128.

17 Ibid.

18 Letter to J. H. Hooker, 13 May 1848, *Correspondence* 4: p. 140.

19 See Martha Richmond (1988) 'Darwin's Study of the Cirripedia', Appendix II of *Correspondence* 4: pp. 391–2.

20 The four von Baer rules of development have been summarized as the following:
    (i) The more general characters of a large group of animals appear earlier in their embryos than the more special characters.
    (ii) From the most general forms the less general are developed, and so on, until finally the most special arise.
    (iii) Every embryo of a given animal form, instead of passing through the other forms, becomes separate from them.
    (iv) Fundamentally, therefore, the embryo of a higher form never resembles any other form, but only its embryo. From Rudolf A. Raff and Thomas C. Kaufman, *Embryos, Genes and Evolution: The Developmental-Genetic Basis of Evolutionary Change* (New York: Macmillan Publishing Co., 1983), p. 9.

21 For a closer examination of von Baer's ideas see Dov Ospovat, *The Development of Darwin's Theory: Natural History, Natural Theology and Natural Selection, 1838–1859* (Cambridge: Cambridge University Press, 1981), pp. 117–24.

22 Henri Milne-Edwards, 'Considérations sur Quelques Principes Relatifs à la Classification Naturelle des Animaux', *Annales des Sciences Naturelles*, 3rd series, vol. 1 (1844).

23 Ibid., p. 66; see also Dov Ospovat, op. cit., p. 124–9.

24 Charles Darwin, *A Monograph on the Sub-Class Cirripedia: The Balanidae* (London: Ray Society, 1854), p. 23.

25 Letter to Joseph Hooker, 6 October 1848, *Correspondence* 4: p. 169.

26 CD to Emma Darwin, 28 May 1848, *Correspondence* 4: p. 147.

27 Darwin kept detailed accounts of the books he read. The details here about the books he read in 1848 and his reactions to them are to be found in 'Darwin's Reading Notebooks', Appendix IV of *Correspondence* 4: pp. 476–7.

28 J. A. S. Grenville, *Europe Reshaped 1848-1878* (London: Fontana, 1976),
    pp. 80-83.

29 F. Engels and K. Marx, *The Communist Manifesto* (1848; Oxford:
    Oxford University Press, 1848; 1992).

30 Later Marx was to quote enthusiastically a reviewer who had observed that
    'The scientific value of such an inquiry [*The Communist Manifesto*] lies in the
    disclosing of the special laws that regulate the origin, existence, development,
    death or a given social organism and its replacement, by another higher one.'

31 DAR 72:117-19v.

32 Louis Agassiz had been invited to give a course of public lectures in Boston in
    the United States by John Amory Lowell in 1846. These lectures, called 'Plan
    of Creation in the Animal Kingdom', were so popular that Agassiz decided to
    remain in the United States; E. Lurie, *Louis Agassiz: A Life in Science*
    (Chicago: University of Chicago Press, 1960), pp. 116-33.

33 Letter to Jean Louis Rodolphe Agassiz, 22 October 1848, *Correspondence* 4:
    p. 180.

34 Darwin was wrong about the cement glands – see William A. Newman (1993),
    'Darwin and Cirripedology', in *Crustacean Issues*, vol. 8, pp. 368-70.

35 Ibid., p. 179.

36 Ibid., pp. 178-80.

37 *Autobiography*, pp. 86-7.

6 Very Like a Lobster

1 This fictionalized account is based upon two primary sources. The first is
   Darwin's account of the Water Cure treatment tailored for him by Gully in his
   letter to his sister Susan on 19 March 1849, *Correspondence* 4: pp. 224-5.
   Darwin called this his 'hydropathical diary'. The second source is Joseph
   Leech's comic and detailed account of his three weeks at Malvern in 1850,
   which was anonymously published as Joseph Leech, *Three Weeks in Wet
   Sheets; Being the Diary and Doings of A Moist Visitor to Malvern* (London:
   Hamilton, Adams & Co; Bristol: John Ridler; Malvern: Lamb and Son, 1856).

2 Joseph Hooker, *Himalayan Journals; or notes of a Naturalist in Bengal, the
   Sikkim and Nepal Himalayas, the Khasia Mountains, &c.* 2 vols (London.:
   John Murray, 1854), vol. 1, p. 115.

3 Letter from J. D. Hooker to Darwin, 3 February 1849, *Correspondence* 4: p. 195.

4 George's recollections of Malvern in DAR 112.

5 James Gully, *The Water Cure in Chronic Disease* (London: John Churchill;
   Malvern: Henry Lamb, 1846), p. 515.

6 Tennyson's poems of this period, particularly 'The Princess' (1847) and 'In
   Memoriam' (1850), are full of evolutionary material drawn from Tennyson's
   reading of Lyell's *Principles of Geology* and Robert Chambers' *Vestiges of the
   Natural History of Creation*.

7   James Gully, op. cit., p. 550.

8   Ibid., p. 655.

9   Ibid., p. 658.

10 Cora Weaver, *Malvern as a Spar Town* (Malvern: Cora Weaver Press, 1991), p. 43.

11 Joseph Leech, op. cit., pp. 37–8.

12 CD to Susan Darwin, 19 March 1849, with additional remarks added by Emma Darwin, *Correspondence* 4: p. 224.

13 Joseph Leech, op. cit., p. 92.

14 James Gully, op. cit., p. 659.

15 Gully charged three guineas a week to his out-patients, with an additional weekly payment of four shillings to the Bath Man. The first consultation with Gully cost two guineas.

16 Thomas Carlyle to R. W. Emerson, cited in E. S. Turner, *Taking the Cure* (London: Michael Joseph, 1967), p. 175.

17 Henrietta Litchfield, ed., *Emma Darwin: A Century of Family Letters 1792–1896*, 2 vols (London: John Murray, 1915), vol. 2, pp. 122–3.

18 CD to Susan Darwin, 19 March 1849, with additional remarks added by Emma Darwin, *Correspondence* 4: p. 225.

19 Joseph Leech, op. cit., pp. 75–6.

20 Ibid., p. 43.

21 CD to Hugh Strickland, 4 February 1849, *Correspondence* 4: p. 206.

22 From Hugh Strickland's abstract of Charles Darwin's paper objecting to existing systems of nomenclature, preserved in the University Museum of Zoology, Cambridge, and printed in *Correspondence* 4: pp. 188–9.

23 CD to Hugh Strickland, 4 February 1848, *Correspondence* 4: p. 206.

24 J. D. Hooker to Charles Darwin, 3 February 1849, *Correspondence* 4: p. 204.

25 CD to Hugh Strickland, 4 February 1849, *Correspondence* 4: p. 207.

26 CD to Syms Covington, 30 March 1849, *Correspondence* 4: p. 230.

27 Joseph Leech, op. cit., p. 103.

28 George Darwin's account of Malvern, DAR 112: p. 49.

29 CD to Charles Lyell, 14 June 1849, *Correspondence* 4: p. 239.

30 CD to Henslow, 6 May 1849, *Correspondence* 4: p. 236.

31 Letter from J. D. Hooker to Darwin, 24 June 1849, Camp Sikkim, Himalayas, *Correspondence* 4: p. 245.

32 'A Visit to Darwin's Village – Reminiscences of Some of his Humble Friends from a Special Correspondent', *Evening News* (London), Friday, 12 February 1909, p. 4.

33 Ibid..

34 See note to letter written by CD to J. G. Forchhammer, 25 September 1849, *Correspondence* 4: p. 256.

35 Letter from Darwin to J. . Forchhammer, 1 December 1849, *Correspondence* 4: p. 282–3.

36 CD to Charles Lyell, 2 September 1849, *Correspondence* 4: p. 271.

37 CD to Albany Hancock, 25 December 1849, *Correspondence* 4: pp .291–2.
38 DAR 112: pp .9–49.
39 CD to J. D. Hooker, 3 February 1850, *Correspondence* 4: p. 310–11.
40 Ibid., p. 311.

7 On Speculating

1 'Darwin [Erasmus] possesses the *epidermis* of poetry but not the *cutis*; the
    cortex without the *liber*, *alburnum*, *lignum*, or *medulla*.' Samuel Taylor
    Coleridge on *The Botanic Garden*, in *Anima Poetae* (originally published
    1805; London: Heinemann, 1895), p. 280.
2 See Janet Browne, *Charles Darwin: Voyaging* (London: Pimlico, 1996),
    pp. 7–10. See also William Darwin's recollections of his father's financial
    advice in DAR112, 3c.
3 See pages of *The Times,* in particular between 25 October and December
    1845. There were twenty-nine articles in *The Times* on railway speculation
    from June 1845 to December 1845.
4 See Jonathan Smith, *Fact and Feeling: Baconian Science and the Nineteenth-
    Century Literary Imagination* (Madison and London: University of
    Wisconsin Press, 1994), and Alborn, Timothy L., 'The business of induction:
    industry and genius in the language of British scientific reform, 1820–1840',
    History of Science vol. 34 (1996), pp. 91–121.
5 William Whewell's review of *The Works of Francis Bacon*, ed. James
    Spedding, Robert Leslie Ellis and Douglas Denon Heath, *Edinburgh Review*,
    vol. 106 (1857), pp. 314–15.
6 See Stephen J. Gould, 'Introduction' to *The Mismeasure of Man* (New York:
    Norton, 1981); also Donna Haraway, *Primate Visions* (New York: Routledge,
    1989), Introduction and chapter 14, for interesting discussions of this point.
7 Charles Darwin, *The Autobiography of Charles Darwin 1809–1882*, edited and
    with appendix and notes by Nora Barlow (London: Collins, 1958), p. 121.
8 Later CD claimed that he had never properly acknowledged his debt to Lyell,
    writing 'when seeing a thing never seen by Lyell, one yet saw it partially
    through his eyes'. Francis Darwin, *Life and Letters* (New York, 1888) vol. 2, p.
    55; cited Jonathan Smith, 'Seeing Through Lyell's Eyes: The Uniformitarian
    Imagination and *The Voyage of the Beagle*', in *Fact and Feeling: Baconian
    Science and the Nineteenth-Century Literary Imagination* (Madison and
    London: The University of Wisconsin Press, 1994), p. 95.
9 CD to J. D. Dana, 5 December 1849, *Correspondence* 4; p. 286.
10 CD gives a long and detailed account of his writing processes and frustrations
    in 'Darwin's Thoughts on his Mental Processes', in Charles Darwin, *The
    Autobiography of Charles Darwin 1809–1882*, edited, with appendix and
    notes, by Nora Barlow (London: Collins, 1958), pp. 135–45.
11 CD to Charles Lyell, 7 December 1849, *Correspondence* 4: p. 289.

12 Charles Darwin, *The Autobiography of Charles Darwin 1809–1882*, edited, with appendix and notes, by Nora Barlow (London: Collins, 1958), pp. 135–45.

13 See DAR 185.10: 10 in the Darwin collection in Cambridge University Library for the children's drawings on the back of Darwin's manuscripts.

14 Letter to J. D. Dana, 24 February 1850, *Correspondence* 4: p. 313.

15 Charles Darwin, *A Monograph of the Sub-Class Cirripedia* (London Ray Society, 1851), pp. 292–3.

16 CD to Charles Henry Lardner Woodd, 4 March 1850, *Correspondence* 4: pp. 316–7. Darwin's fingers had been burnt by the general discrediting of his Glen Roy hypothesis in the 1840s; he had also seen what had happened to the *Vestiges of the Natural History of Creation*, denounced as a mere theory, unsubstantiated by facts.

17 William Harvey (1849), *The Sea-Side Book: Being an Introduction to the Natural History of the British Coasts* (London: John Van Voorst, 1849), pp. 4–5.

18 Letter from J. D. Hooker to CD, 6 and 7 April 1850, *Correspondence* 4: pp. 327–8.

19 Ibid., pp. 327–30.

20 Darwin's sons George and Francis both comment on Darwin's respect for and anxiety about time. See in particular *The Complete Edited Manuscript of Francis Darwin's Preliminary Draft of the Reminiscences of My Father's Life*, ed. Robert Brown, Independent Study (Manuscript kept in the Darwin Correspondence Project office, Cambridge University Library), p. 54.

21 CD to J. D. Hooker, 13 June 1850, *Correspondence* 4: p. 344.

22 CD to James Sowerby, 13 April 1850, *Correspondence* 4: p. 332.

23 CD to James Sowerby, [12 or 19 August] 1850, *Correspondence* 4: p. 349.

24 CD to Richard Owen, 28 April 1850, *Correspondence* 4: p. 334.

25 CD to W. D. Fox, 10 October 1850, *Correspondence* 4: p. 362.

26 CD to Syms Covington, 23 November 1850, *Correspondence* 4: p. 369.

27 Elizabeth Gaskell, *The Life of Charlotte Bronte* (Heinemann, Penguin, 1997), chapter 18.

28 Lyell's presidential address delivered at the anniversary meeting of the Geological Society on 15 February 1850, p.lxvi. *Correspondence* 4: p. 616; 'Anniversary Address of the President', *Quarterly Journal of the Geological Society of London*, vol. 6, p. lxvi.

29 CD to Charles Lyell, 8 March 1850, *Correspondence* 4; p. 319.

30 Randal Keynes, *Annie's Box: Charles Darwin, His Daughter and Human Evolution* (London: Fourth Estate, 2001), p. 201.

31 Ibid., p. 149.

32 Ibid., p. 152.

33 Ibid., p. 157.

34 Ibid., p. 155.

8 Writing Annie

1 Charles Darwin, A *Monograph of the Sub-Class Cirripedia* (Palaeontographical Society, 1851), pp. 2–3.

2 William Howitt, *The Book of the Seasons; or the Calendar of Nature* (London: Henry Colburn and Richard Bentley, 1831), pp. 16–17. Charles Darwin borrowed this from the London Library in January for Annie.

3 Randal Keynes, *Annie's Box: Charles Darwin, his Daughter and Human Evolution* (London: Fourth Estate, 2001), p. 159.

4 For further interpretations of Darwin's struggle with his religious beliefs see James Moore, 'Of Love and Death: Why Darwin Gave up Christianity', in J. Moore, ed., *History, Humanity and Evolution: Essays for John C. Greene* (Cambridge: CUP, 1989), pp. 95–229.

5 See Darwin's Reading Notebook DAR 119:22b.

6 Harriet Martineau, *Eastern Life: Present and Past*, 3 vols (London: Edward Moxon, 1848), p. 334.

7 H. G. Atkinson and H. Martineau, *Letters on the Laws of Man's Nature and Development* (London, John Chapman, 1850).

8 Ibid., p. 10.

9 Randal Keynes, op. cit., p. 169.

10 CD to Emma Darwin 21 April 1851, *Correspondence* 5: p. 20.

11 CD to Emma Darwin, 18 April, *Correspondence* 5: p. 14.

12 CD to Emma Darwin, 17 April 1851, *Correspondence* 5: p. 13.

13 Emma Darwin to CD, 19 April, *Correspondence* 5: p. 15.

14 Cited Randal Keynes, op. cit., p. 172.

15 CD to Emma Darwin, 19 April 1851, *Correspondence* 5: p. 16.

16 Fanny wrote to her daughter Effie that it was 'just at twelve o'clock that we heard her breathe for the last time, while the peals of thunders were sounding. Poor Miss Thorley was very ill after and Brodie too. I never saw anyone suffer as she did, but they are both going tomorrow, Brodie to Down and Miss Thorley to her own home. She wants rest.' Cited Keynes, op. cit., p. 176.

17 CD to Emma Darwin, 23 April 1851, *Correspondence* 5: p. 24.

18 Emma Darwin to CD, 24 April 1851, *Correspondence* 5: p. 24.

19 Fanny Wedgwood to CD and ED, 25 April 1851, *Correspondence* 5: p. 29.

20 Charles Darwin, *Living Cirripedia* (London: The Ray Society, 1851), pp. 67–8.

21 CD's account of Annie can be found as Appendix II of *Correspondence* 5, pp. 540–3. It is also reprinted in its entirety in Randal Keynes, op. cit., pp. 195–8.

22 Emma kept a careful record of Etty's questions in the months after Annie died and her own responses to them. These are reproduced in Appendix II of *Correspondence* 5, pp. 542–3.

23 From *The Rubaiyat of Omar Khayyam*, translated by Edward Fitzgerald in 1859 (London: Foulis, 1917).

9 Corked and Bladdered Up

1 CD to J. S. Disnurr, 6 May 1851, *Correspondence* 5: pp. 36–7.

2 CD to Edward Forbes, May–June 1851, *Correspondence* 5: p. 33.

3 One of the ironies of this letter is that subsequent research (D. J. Crisp, 'Extending Darwin's Investigation on the Barnacle Life-History', *Biological Journal of the Linnean Society*, vol. 20 (1983), pp. 73–83) has shown that Darwin made a significant blunder in his interpretation of the female parts of the barnacle in his identification of the oviduct as the cement-making gland.

4 CD to Charles Spence Bate, 13 June 1851, *Correspondence* 5: p. 43; CD's reference here to being pipped to the post by others refers to his own work on the geographical description of Arctic and alpine plants, which Edward Forbes anticipated in 1846.

5 CD to Charles Spence Bate, 13 June 1851, *Correspondence* 5: pp. 43–4.

6 Ibid., p. 44.

7 CD to J. S. Disnurr, 13 June 1851, *Correspondence* 5, p. 45. For Forbes' use of the dredge see Philip F. Rehbock, 'The Early Dredgers: Naturalizing in British Seas, 1830–1850', *Journal of History of Biology*, vol 12 (1979), pp. 293–368.

8 See Adrian Desmond, 'Robert E. Grant: The Social Predicament of a Pre-Darwinian Transmutationist', in *The Journal of the History of Biology*, vol. 17, no. 2 (Summer 1984), pp. 189–223.

9 Darwin was happy to write Huxley a reference in October that year for his application for a chair in natural history in Toronto, Canada. But Huxley would not be successful.

10 CD to Robert Ball, 26 May 1851, *Correspondence* 5: pp. 37–8. He did ask to see a South American barnacle in the collection, however, which he probably thought might be another Arthrobalanus.

11 See *Correspondence* 5, pp. 38, 40, 47.

12 There is a blank in the copy of the letter here where the copyist clearly could not decipher the word.

13 CD to J. S. Disnurr, 13 June 1851, *Correspondence* 5: p. 45.

14 R. D. Keynes, ed., *Charles Darwin's Beagle Diary* (Cambridge: CUP, 1988), p. 49.

15 Roualeyn George Gordon Cumming, *Five Years of a Hunter's Life in the Far Interior of South Africa*, 2 volumes (London: John Murray, 1850), vol. 1, p. 358.

16 Herbert Edwardes, *A Year on the Punjab Frontier in 1848–9*, 2 volumes (London: Richard Bentley, 1851), vol. 1, p. 29.

17 *Works of Darwin* vol. 11: *The Lepadidae*, p. 238.

18 CD to George Newport, 24 July 1851, *Correspondence* 5: p. 51.

19 Ibid.

20 I am grateful to Polly Smith and Dr Nick Evans of the Ray Society for tracking down Edwin Lankester's London address.

21 Jeffrey A. Auerbach, *The Great Exhibition of 1851: A Nation on Display* (New Haven and London: Yale University Press: 1999), p. 178.

22 *The Athenaeum*, no. 1240, 2 August 1851, p. 833. Emma's diary records that they visited 'The Polytechnic Overland Mail' and the 'Polytechnic Conjurors'. There were two diorama/ panoramas that summer called 'The Overland Mail to India'; neither were in the Polytechnic on Regent Street, but the show described by *The Athenaeum* was held in the Gallery of Illustration, also on Regent Street, which probably accounts for Emma's confusion. The Darwins would have seen the Gallery of Illustration show rather than the one held in the Willis Rooms by Albert Smith, for Altick states that Albert Smith's show had gone on tour by the end of July when the Darwins arrived (Altick, p. 474). For further information about the history of the diorama see 'The Diorama in Great Britain in the 1820s', by R. Derek Wood in the *History of Photography*, Autumn 1993, vol. 17, no. 3, pp. 284–95; and Altick, Richard D, *The Shows of London* (Cambridge, Mass., USA : Belknap Press, 1978).

23 *The Athenaeum*, no. 1240, 2 August 1851, pp. 830–1.

24 Ibid.

25 Dickens wrote about the London panoramas in *Household Words*, through a fictional character addicted to panoramas, 'Mr Booley', *Household Words* 1 (1850), 73–7.

26 CD to C. S. Bate, 18 August 1851: 'Smith & Beck of 6 Colman St City, assure me that the Asphalte which they sell in 1s bottles is better than Gold size for the purpose, I mentioned to you, I have got a bottle but have not yet tried it.' *Correspondence* 5: p. 56.

27 CD to George Newport, 24 July 1851, *Correspondence* 5: pp. 51–2, CD to George Newport, 12 August 1851, *Correspondence* 5: pp. 54–5.

28 For an account of the launch of the *Westminster Review* and of the relationship between George Eliot and John Chapman see Gordon S. Haight, *George Eliot and John Chapman with Chapman's Diaries* (London: Archon Books, 1969).

29 CD to Edwin Lankester, 7 August 1851, *Correspondence* 5: pp. 53–4.

30 CD to William Erasmus Darwin

10 Drawing the Line

1 In October 1852 Darwin wrote to his cousin William Fox about the pears he had in his garden and which Fox had recommended ten years before. See CD to W. D. Fox, 24 October 1852, *Correspondence* 5, p. 100.

2 This section is based on Darwin's analysis of pear and apple tree cultivation in *Origin of Species*.

3 Justus von Liebig, *Familiar Letters On Chemistry, In Its Relations To Physiology, Dietics, Agriculture, Commerce, And Political Economy* (3rd ed., London: Taylor, Walton & Maberly, 1851), Letter 11.

4 See CD to J. S. Bowerbank, 28 September 1851, *Correspondence* 5: p. 62.

5 CD to J. J. S. Steenstrup, 16 October 1851, *Correspondence* 5: p. 66.

6 CD continued to write polite and exasperated notes to Steenstrup throughout late 1851.

7 Throughout 1851 and 1852 Darwin sent out his complimentary copies of the *Fossil Stalked Barnacles* (1851) and the *Living Stalked Barnacles* (1851). He particularly targeted those men who who were not members of the Ray Society, who would automatically receive their copy. One man whom Darwin knew would not be able to afford the Ray Society subscription was Thomas Huxley. He sent Huxley a copy of *Living Stalked Barnacles* in 1852 and drew his attention to the *Ibla* and *Scalpellum* discoveries.

8 Joseph Dalton Hooker with Thomas Thomson, *Flora Indica*, Introductory essay (London, 1855) p. 12.

9 See the interesting discussion of this period of Hooker's life in Desmond Ray, *Sir Joseph Hooker: Traveller and Plant Collector* (London: Antique Collectors Club and the Royal Botanical Gardens, Kew, 1999), pp. 200–9.

10 Letter from Hooker to CD, November 1851, *Correspondence* 5: p. 68.

11 In a letter CD wrote to Richard Owen on 17 July 1852 Darwin admitted to having been 'frightened at the thoughts of all the sessile species', so frightened that he had neglected to develop some of his thoughts on the anatomical structures of the stalked living barnacles in volume 2. *Correspondence* 5: p. 95.

12 See letter from CD to J. D. Dana, 8 May 1852, *Correspondence* 5: p. 91: 'I am most vexed at the wooden pill Box with the Crustacean having been lost: I put it in the parcel myself.'

13 CD to W. D. Fox, 7 March 1852, *Correspondence* 5: p. 83.

14 CD to John Stevens Henslow, 11 December 1851, *Correspondence* 5: p. 75.

15 *The Annual Register*, vol. 44, 1852, pp. 478–81.

16 Philip Henry Gosse, *A Naturalist's Sojourn in Jamaica*, assisted by Richard Hill (London: Longman Brown, Green and Longmans, 1851). Darwin lists the book in his reading notebooks.

17 Emma Darwin letter to William Darwin, 23 April 1852: DAR 219.1.4.

18 In 1858, Hooker wrote to Asa Gray that he was 'most thankful . . . that I can now use Darwin's doctrines – hitherto they have been kept secrets I was bound to honour to know, to keep, to discuss with him in private – but never to allude to in public, & I had always in my writings to discuss the subjects of creation, variation &c &c as if I had never heard of Natural Selection – which I have all along known & feel to be not only useful in itself as explaining many facts in variation, but as the most fatal argument against "Special Creation" & for "Derivation" being the rule of all species.' Quoted in Duncan Porter, 'On the road to the *Origin* with Darwin, Hooker, and Gray', *Journal of the History of Biology,* vol. 26 (1993), pp. 33–4.

19 Quoted ibid., p. 13.

20 Hooker later told Francis Darwin: 'It was an established rule that he every day pumped me, as he called it, for half an hour or so after breakfast in his study, when he first brought out a heap of slips with questions botanical, geographical, &c, for me to answer. And concluded by telling me of the progress he had made in his own work, asking my opinion on various points.' Cited in Mea Allen, *The Hookers of Kew, 1785–1911* (London: Michael Joseph, 1967), p. 194. He also remembered: 'long walks, romps with the children on hands and knees, music that haunts me still'. I am indebted to Jim Endersby, of the History and Philosophy of Science Department at Cambridge University, for long discussions about the relationship between Darwin and Hooker and Hooker's struggles to make botany philosophical.

21 CD to Richard Owen, 17 July 1852, *Correspondence* 5, p. 95.

22 CD to William Fox, 24 October 1852, *Correspondence* 5, p. 100.

23 Henry Holland, *Chapters on Mental Physiology* (London: Longman, Brown, Green and Longmans, 1852), Preface, pp. i–x; for an analysis of Holland's place in the development of theories of Victorian psychology see Rick Rylance, *Victorian Psychology and British Culture 1850–1880* (Oxford: Oxford University Press, 2000), pp. 127–43.

24 Henry Holland, *Chapters on Mental Physiology* (London: Longman, Brown, Green and Longmans, 1852), pp. 83–4.

25 Holland (1852), p. 209.

26 CD to William Fox, 24 October 1852, *Correspondence* 5: p. 101.

27 Ibid., p. 100.

28 Ibid., p. 101, note 3.

29 According to Lubbock, CD had 'induced my father to give me a microscope, he let me do drawings for some of his books, and I greatly enjoyed my talks and walks with him. My first scientific original work was on some of his collections.' Cited in Horace Gordon Hutchinson, *Life of Sir John Lubbock, Lord Avebury*, 2 vols (London: Macmillan, 1914), vol. 1, p. 23.

30 John Lubbock, *Natural History Notebook*, 1852–1855; Royal Society LUA.1 Location xxvii. d–e.

31 John Lubbock published several infusoria and crustacean articles in 1853: 'On Labidocera', *Annals and Magazine of Natural History*, vol. xi (1853), and several articles on Calanidae in the *Annals and Magazine of Natural History*, vols xii, lxvii and lxix.

32 CD to J. D. Dana, 25 November 1852, *Correspondence* 5: p. 103.

33 CD to Albany Hancock, 25 December 1852, *Correspondence* 5: pp. 106–8.

34 CD to Albany Hancock, 10 January 1853, *Correspondence* 5, p. 111.

35 CD to Albany Hancock, 12 February 1853, *Correspondence* 5: p. 116.

36 CD to W. D. Fox, 29 January 1853, *Correspondence* 5: p. 113.

37 CD to Albany Hancock, 12 February 1853, *Correspondence* 5: pp. 116–7.

38 CD to William Darwin, 1 March, 1853, Correspondence 5: p. 121.

39 See *Annual Register* vol. 95, 1853, pp. 32–3.

40 Ibid., pp. 28–32.

41 CD to Albany Hancock, 29 January 1853, *Correspondence* 5: p. 114.

42 Albany Hancock to CD, 25 February 1853, *Correspondence* 5: p. 121.

43 CD to Edwin Lankester, Ray Society, 19 March 1853, *Correspondence* 5: p. 123.

11 Manoeuvres and Skirmishes

1 Darwin kept a microscope slide catalogue, which is now in the Cambridge Zoology Museum, *History of the Collection*, vol. III, 1892–97, no. 454. The microscope catalogue that lists the 11 Arthrobalanus slides shows that CD retained the Arthrobalanus name right to the end of his research, for at the top of the page he writes: 'Arthrobalanus = *Cryptophialus*'.

2 Philip Henry Gosse, *A Naturalist's Sojourn in Jamaica*, assisted by Richard Hill (London: Longman, Brown, Green and Longmans, 1851), p. v.

3 CD to Albany Hancock, 30 March 1853, *Correspondence* 5: p. 126. Subsequent zoological research has shown that *Alcippe* (now called *Trypetesa*) and *Cryptophialus* have much closer affinities than Darwin believed. See William A. Newman, 'Darwin and Cirripedology', in *Crustacean Issues*, vol. 8 (Rotterdam: A. A. Balkema, 1993), pp. 385–6. I am grateful to Shelley Innes of the Darwin Correspondence Project for drawing Newman's work to my attention.

4 CD to Albany Hancock, 30 March 1853, *Correspondence* 5: p. 127.

5 Charles Darwin, *A Monograph of the Sub-Class Cirripedia, Vol. II: The Balanidae,* vol. 12 of *The Complete Works of Charles Darwin*, ed. Paul H. Barrett & R. B. Freeman (London: William Pickering, 1988), p. 26.

6 Ibid., p. 29.

7 Ibid., p. 500.

8 Ibid., p. 501.

9 *Annual Register*, vol. 95 (1853), p. 48.

10 CD to T. H. Huxley, 11 April 1853, *Correspondence* 5: p. 130.

11 See Adrian Desmond, *Huxley: From Devil's Disciple to Evolution's High Priest* (Harmondsworth: Penguin, 1997), p. 178. Huxley wrote in the after math of his mother's death: 'Belief and Happiness seem to be beyond the reach of thinking men in these days but Courage and Silence are left.' Huxley's mother died on 14 April; she was buried in Barming on 19 April. CD reported having read Huxley's mollusc paper in a letter of 23 April, which means that Huxley must have sent it between receiving Darwin's first letter dated 11 April, which would have arrived on 12 or 13 April. The most likely date therefore that Huxley sent the mollusc paper, which he would have had to copy out by hand before sending, would have been in the two or three days after his mother died.

12 T. H. Huxley, 'On the Morphology of the Cephalous Mollusca, as illustrated

by the anatomy of certain Heteropoda and Pteropoda collected during the voyage of the H.M.S. *Rattlesnake* in 1846–50, *Philosophical Transactions of the Royal Society of London* vol. 143 (1853), pp. 29–65.

13 CH to T. H. Huxley, 23 April 1853, *Correspondence* 5: p. 133.

14 Ibid., p. 133–4.

15 CD had first read Liebig's book in November 1844 (*Correspondence* 4: p. 469); he read it again in October 1851, CD to William Darwin, 3 October 1851, *Correspondence* 5: p. 63. I owe a debt of gratitude to Randal Keynes, who is currently writing a book on the garden of Down House, for talking through the history of the tank building and sewage works with me and for drawing my attention to new sources of information.

16 For an intriguing interpretation of Chadwick's schemes and their place in nineteenth-century culture see David Trotter, *Circulation: Defoe, Dickens and the Economies of the Novel* (Basingstoke: Macmillan, 1988).

17 CD to E. A. Darwin, 26 April 1853, *Correspondence* 5: p. 136.

18 CD to Susan Darwin (CD's sister, who had a special interest in Darwin's live stock), 27–8 April 1843: 'You ask also about the liquid manure; part of my apparatus is complete & the tank is made & is now beginning to fill: I hope to try one cask this spring.' *Correspondence* 2: p. 360.

19 CD to *Gardeners' Chronicle*, c.27 April 1853, *Correspondence* 5: p. 138.

20 Gutta percha was derived from trees growing around the Pacific Basin. It became flexible when heated, yet rock-hard when cool. It had been one of the star new products of the Great Exhibition. The London gutta percha factory burned down in June that year. An accidental fire took hold of the highly inflammable stocks and the fire would have caused massive destruction across this part of London if the fire service had not intervened. *Annual Register*, vol. 95 (1853), p. 68.

21 CD paid £32/3s to the builder and bricklayer Issac Laslett on 22 April, and a second instalment of £40/15s on 4 July. The pipes cost him £1/7s and £1/15s in two payments in June and August. See *Correspondence* 5: p. 140, note 1. I am grateful to Randal Keynes for supplying me with the following information about water tanks from George William Johnson, F.R.H.S. *The Cottage Gardener's Dictionary; describing the plants, fruits & vegetables desirable for the garden, etc* (London: 1852), which claims that the best water 'for the gardener's purposes is rain water, preserved in tanks sunk in the earth, and rendered tight either by puddling, or bricks covered with Parker's cement. To keep these tanks replenished, gutters should run round the eaves of every structure in the garden and communicate with them.' The water from the well in the chalk was presumably hard and therefore, in the words of the Dictionary, 'invariably prejudicial' to plants.

22 CD to W. D. Fox, 17 July 1853, *Correspondence* 5: p. 147.

23 Darwin's account book registers a final payment to the builders for 4 July. The invoice for the last instalment of gutta percha pipes is dated 18 August.

24 CD to J. A. H. de Bosquet, 18 June 1853, *Correspondence* 5: p. 143.

25 Now Marine Parade, Eastbourne.

26 Godfrey Charles Munday, *Our Antipodes; or, Residence and Rambles in the Australasian colonies. With a glimpse of the Gold Fields* (London: Richard Bentley, 1852), vol. 3, p. 385.

27 CD to W. D. Fox, 17 July 1853, *Correspondence* 5: p. 147.

28 'Darwin's Observations on his Children', Appendix 3 in *Correspondence* 4: p. 426.

29 This account is based on descriptions of the Darwin children from Darwin's own observations, reprinted as Appendix 3: 'Darwin's Observations on His Children,' *Correspondence* 4: pp. 410–33, and on the accounts given in Randal Keynes, *Annie's Box* (London, Fourth Estate, 2000).

30 Francis Galton, *The Narrative of An Explorer in Tropical South Africa* (London: John Murray, 1853), p. v. Darwin lists this book in his reading notebooks.

31 CD to Francis Galton, 24 July 1853, *Correspondence* 5: p. 149–50.

32 CD to W. D. Fox, 29 July 1853, *Correspondence* 5: p. 151.

33 CD to W. D. Fox, 10 August 1853, *Correspondence* 5: p. 151.

34 Etty described the Chobham visit later: 'Our visit was planned in order to see what we could of the camp with its mimic warfare. I well remember my father's intense enjoyment of the whole experience. Admiral Sulivan, his old shipmate on board the *Beagle*, showed us about and greatly added to our pleasure.' Henrietta Litchfield, ed., *Emma Darwin: A Century of Family Letters* (London: John Murray, 1915), vol. 2, p. 154. My description of Chobham Camp is based upon an eyewitness account produced in *Annual Register*, vol. 95 (1853), pp. 78–9.

35 In the Preface to *Living Cirripedia* (1854) Darwin wrote: 'I had originally intended to have published a small volume on my anatomical observations; but the full abstract given in my former volume, which will be illustrated to a certain extent in the plates appended to this volume, together with the observations here given under the Balanidae, appear to me sufficient, and I am unwilling to spend more time on the subject.' I am grateful to Shelley Innes of the Darwin Correspondence Project for this information.

36 [George Henry Lewes] 'Lyell and Owen on Development', *Leader*, 18 October 1851, p. 996.

37 For an account of the changes in the intellectual liberal debates in London in the early 1850s see Jim Secord, *Victorian Sensation: The Extraordinary Publication, Reception and Secret Authorship of 'Vestiges of the Natural History of Creation'* (Chicago: Chicago University Press, 2000), pp. 479–93.

38 Although there may have been letters that have not survived, the surviving correspondence suggests that the two friends exchanged no letters between April 1852 and September 1853. See *Correspondence* 5.

39 Joseph Dalton Hooker, *Flora Novae-Zelandiae* (London: Lovell Reeve, 1853–5), Introductory essay, p. viii.

40 William Henslow Hooker was born on 25 January 1853; see CD to W. D. Fox, 29 February 1853, *Correspondence* 5: p. 113.

41 CD to J. D. Hooker, 25 September 1853, *Correspondence* 5: p. 155.

42 I am grateful for Jim Endersby's comments on Hooker's difficulties with species arguments and taxonomy at this point in his career.

43 Joseph Dalton Hooker, op. cit., Introductory essay, p. xii.

44 CD to J. D. Hooker, 25 September 1853, *Correspondence* 5: p. 155–6.

45 Ibid., p. 156.

46 CD to J. D. Hooker, 9 October 1853, *Correspondence* 5: pp. 158–9.

47 CD to J. D. Hooker, 10 October 1853, *Correspondence* 5: p. 161.

48 See note 32 above.

49 CD to J. D. Hooker, 5 November 1853, *Correspondence* 5: p. 166.

50 CD to Palaeontographical Society [before 24 February 1854], *Correspondence* 5: p. 177.

51 J. D. Hooker to CD, 26 February 1854, *Correspondence* 5: p. 178.

52 CD to J. D. Hooker, 26 March 1854, *Correspondence* 5: p. 187.

53 J. D. Hooker to CD, 29 June 1854, *Correspondence* 5: p. 199.

54 CD to J. D. Hooker, 7 September 1854, *Correspondence* 5: p. 214–15.

12 The Universe in a Barnacle Shell

1 Francis recalls his father's brightly coloured dressing gown in *The Completed Edited Manuscript of Francis Darwin's Preliminary Draft of the 'Reminiscences of My Father's Everyday Life'*, ed. Robert Brown, Independent Study. Manuscript in the Darwin Correspondence Project office.

2 DAR 113. I am grateful to Shelley Innes of the Darwin Correspondence Project for this information.

3 Emma Darwin's diary. John Lubbock writes in his diary for October 1854: 'Dined at Darwins to meet Lyell and Hooker', John Lubbock, Diary for 1853–63, British Library Add 62679. I am grateful to Randal Keynes for this information.

4 In the *Origin*, Darwin concluded that 'no clear line of demarcation has as yet been drawn between species and subspecies – that is, the forms which in the opinion of some naturalists come very near to, but do not quite arrive at the rank of species; or, again, between sub-species and well-marked varieties, or between lesser varieties and individual differences. These differences blend into each other in an insensible series; and a series impresses the mind with the idea of an actual passage ... Hence I look at individual differences, though of no small interest to the systematist, as of high importance for us, as being the first step towards slight varieties ... And I look at varieties which are in any degree more distinct and permanent, as steps leading to more strongly-

marked and permanent varieties; and at the latter, as leading to sub-species, and to species . . . Hence I believe a well-marked variety may be called an incipient species. Charles Darwin, *On the Origin of Species by Natural Selection* (London: John Murray, 1859), p. 53.

5 For further studies on the influence and impact of Darwin's barnacle work on his later work, see A. J. Southward, 'A New Look at Variation in Darwin's Species of Acorn Barnacles', *Biological Journal of the Linaean Society*, vol. 20 (1983), pp. 59–72; Ghiselin, M. T. (1969), *The Triumph of the Darwinian Method* (California: California University Press, 1969); Ghiselin, M. T. & Jaffe, L. 'Phylogenetic Classification in Darwin's monograph on the subclass Cirripedia', *Systematic Zoology*, vol. 22 (1973), pp. 132–40; D. J. Crisp, 'Extending Darwin's Investigations on the Barnacle Life-History', *Biological Journal of the Linnaean Society*, vol. 20 (1983), pp. 73–83; Martha Richmond, 'Darwin's Study of the Cirripedia', in *Correspondence* 4: pp. 388–409.

6 For a history of this book's publication and reception see James A. Secord, *Victorian Sensation: The Extraordinary Publication, Reception, and Secret Authorship of 'Vestiges of the Natural History of Creation'*, (Chicago and London: Chicago University Press, 2000).

7 Thomas Henry Huxley, review of *Vestiges of the Natural History of Creation*, Tenth Edition, London 1853, in *The British and Foreign Medico-Chirurgical Review* (1854), *Scientific Memoirs* 5.

8 CD to Thomas Henry Huxley, 2 September 1854, *Correspondence* 5: p. 213.

9 Thomas Henry Huxley, 'Science', in *Westminster Review*, vol. 61 (1854), pp. 264–5.

10 Edward Forbes, 'Shell-Fish: Their Ways and Works', *Westminster Review*, vol. 57 (1 January 1852), pp. 42; Forbes died in November 1854 from an abscess on his kidney.

11 Philip Henry Gosse, *A Naturalist's Rambles on the Devonshire Coast* (London, John Van Voorst, 1853).

12 John Timbs, *Curiosities of London: Exhibiting the Most Rare and Remarkable Objects of Interest in the Metropolis with Nearly Fifty Years' Personal Recollections* (London: David Bogue, 1855), p. 780. For further information about the history of the vivarium in the Zoological Gardens see Leonard Robert Brightwell, *The Zoo Story* (London: Museum Press: London, 1952); John Edwards, *London Zoo from old photographs 1852–1914* (London: 26 Rhondda Grove, E3 5AP.); Philip Arthur Richard Street, *The London Zoo* (London: Odhams Press, 1956). For a history of the aquarium generally see Philip F. Rehbock, 'The Victorian Aquarium in Ecological and Social Perspective', in M. Sears and D. Merriman, eds, *Oceanography: The Past* (New York: Springer-Verlag, 1980).

13 J. D. Hooker, [before 17 March 1855], *Correspondence* 5: p. 285.

14 CD to J. S. Henslow, 26 March 1855, *Correspondence* 5: p. 292.

15 CD wrote to Henslow on 13 March to request the Anacharis; it arrived by post on 26 March; *Correspondence* 5: pp. 283, 292. For further information on Anna Thynne, one of the first aquarium owners in London, who kept madrepores alive in glass tanks in Westminster Abbey, see my forthcoming Short Life biography *Anna Thynne* (Short Books, 2002). Thynne was one of the first to discover that a tank containing seawater and seaweed would oxygenate. See also Charles Kingsley's description of how to build a freshwater aquarium in *Glaucus, or the Wonders of the Shore* (Cambridge: Macmillan, 1855), pp. 158–9.

16 See CD's report on the seawater experiments to *Gardeners' Chronicle*, 21 May 1855, *Correspondence* 5: pp. 331–4. Darwin obtained the chemical mixture from the chemist recommended by Philip Gosse, Mr Bolton. Gosse reco mends this chemist in the *Handbook to the Marine Aquarium*: 'The salts are sold in packets, with all needful directions, by Mr Bolton, a chymist [sic] in Holborn.', Philip Gosse, *A Handbook to the Marine Aquarium: containing Instructions for constructing, stocking, and maintaining a tank, and for collecting plants and animals* (London, John Van Voorst, 1855) p. 18

17 CD to J. D. Hooker, 24 April 1855, *Correspondence* 5: p. 320.

18 CD described these experiments in letters to Hooker in March and April 1854: *Correspondence* 5: pp. 299, 305, 308.

19 CD to *Gardeners' Chronicle*, before 26 May 1855, *Correspondence* 5: pp. 337–8.

20 CD to M. J. Berkeley, 12 June 1855, *Correspondence* 5: p. 353.

21 CD to W. D. Fox, 27 March 1855, *Correspondence* 5: p. 294.

22 For an account of Philip Gosse's struggle with the doctrine of natural selection and his response to it in devising his own omphalos theory, see his son's moving book, Edmund Gosse, *Father and Son* (1907; Harmondsworth: Penguin, 1983). See also the excellent recent biography by Ann Thwaite, *Glimpses of the Wonderful: The Life of Henry Philip Gosse*, London: Faber and Faber, 2002.

Epigraph: The Asphalt Curtain

1 Randal Keynes, Darwin's great-grandson, tells me that his grandmother always said that Darwin's hands were distinctively long. His son, Francis, described him as a daddy-long-legs.

2 *Living Cirripedia* (1854), p. 577.

Index

*

Location references in **bold** type refer to illustrations. All barnacle references will be found under 'barnacles'.